A (Terse) Introduction to Linear Algebra

STUDENT MATHEMATICAL LIBRARY
Volume 44

A (Terse) Introduction to Linear Algebra

Yitzhak Katznelson
Yonatan R. Katznelson

AMERICAN MATHEMATICAL SOCIETY
Providence, Rhode Island

2010 *Mathematics Subject Classification.* Primary 15–01.

The cover art is created by Noah Katznelson
and used with permission.

For additional information and updates on this book, visit
www.ams.org/bookpages/stml-44

Library of Congress Cataloging-in-Publication Data

Katznelson, Yitzhak, 1934–
   A (terse) introduction to linear algebra / Yitzhak Katznelson, Yonatan R. Katznelson.
      p. cm. — (Student mathematical library, ISSN 1520-9121 ; v. 44)
   Includes index.
   ISBN 978-0-8218-4419-9 (alk. paper)
   1. Algebras, Linear. I. Katznelson, Yonatan R., 1961– II. Title. III. Title: Introduction to linear algebra.

QA184.2.K38   2008
512′.5—dc22   2007060571

Contents

Contents

Preface

This book is a presentation of the elements of Linear Algebra that every mathematician, and everyone who uses mathematics, should know. It covers the core material, from the basic notion of a finite-dimensional vector space over a general field, to the canonical forms of linear operators and their matrices, obtained by the decomposition of a general linear system into the direct sum of cyclic systems. Along the way it covers such key topics as: systems of linear equations, linear operators and matrices, determinants, duality, inner products and the spectral theory of operators on inner-product spaces. We conclude with a selection of additional topics, indicating some of the directions in which the core material can be applied and developed.

In its mathematical prerequisites the book is elementary, in the sense that no previous knowledge of linear algebra is assumed. It is self-contained, and includes an appendix that provides all the necessary background material: the *very basic* properties of groups, rings, and of the algebra of polynomials over a field. The book is intended, however, for readers with some mathematical maturity and readiness to deal with abstraction and formal reasoning. It is appropriate for an advanced undergraduate course.

As the title implies, the style of the book is somewhat terse. We mean this in two senses.

First, we focus with few digressions on the principal ideas and results of linear algebra *qua* linear algebra. The book contains fewer routine numerical examples than do many other texts, and offers almost no interspersed applications to other fields; these should be adapted to the readership and, if the book is used in a course, provided by the teacher.

Second, the writing itself tends to be concise and to the point, to the extent that some of the proofs might be better described as detailed lists of hints. This is intentional—we believe that students learn more by having to fill in some details themselves.

Besides its style, this book differs from many other texts on the subject in that we try to present the main ideas, whenever possible, in the context of vector spaces over a general field, $\mathbb{F}$, rather than assuming the underlying field to be $\mathbb{R}$ or $\mathbb{C}$. Inner-product spaces, along with the naturally associated classes of self-adjoint, normal, and unitary (or orthogonal) operators, are introduced later than in many books, and the *spectral theorems* for these operators, besides being fundamentally important on their own, also serve here to pave the way for the notions of *reducing* and *semisimplicity* and, eventually, to the general structure theorems—the Jordan form, when the underlying field is algebraically closed, and the corresponding form over general fields.

The text consists of eight chapters and an appendix. These are divided into sections, and further into subsections. Definitions, propositions, examples, etc., are numbered according to the subsection in which they appear, and no subsection has more than one object (definition, theorem, etc.) of each kind. For example, Lemma 1.3.5 is the lemma appearing in subsection 1.3.5, and Theorem 1.3.5 is the theorem appearing in the same subsection. References to the appendix have the form A.x.y (for subsection y of section x, in the appendix).

Exercises appear at the end of sections, and are numbered accordingly, e.g., exercise **ex3.1.2** is the second exercise of section 3.1.

Starred sections, subsections, and exercises contain material that can be skipped on first reading. Several of these sections, as well as parts of the additional topics (in Chapter 8), require some familiarity with basic analysis, e.g., concepts like convergence and continuity.

Chapter 1

Vector Spaces

1.1 Groups and fields

Vector spaces are defined over fields, and the definitions of both fields and vector spaces depend on the notion of a commutative group. Groups and fields are reviewed in more detail in the appendix but, for convenience, we include here their definitions along with some basic examples.

1.1.1 Groups.

DEFINITION: A *group* is a pair $(G, *)$, where G is a set and $*$ is a binary operation $(x, y) \mapsto x * y$, defined for all pairs $(x, y) \in G \times G$, taking values in G, and satisfying the following conditions:

G-1 The operation is associative: For all $x, y, z \in G$,

$$(x * y) * z = x * (y * z).$$

G-2 There exists a unique element $e \in G$, called the *identity element* or the *unit* of G, such that $e * x = x * e = x$ for all $x \in G$.

G-3 For every $x \in G$ there exists a unique element x^{-1}, called *the inverse* of x, such that $x^{-1} * x = x * x^{-1} = e$.

A group $(G, *)$ is *abelian*, or *commutative*, if $x * y = y * x$ for all x and y. The group operation in a commutative group is often written and referred to as *addition*, in which case the identity element is written as 0, and the inverse of x as $-x$.

When the group operation is written as multiplication, the operation symbol $*$ is typically written as a dot (i.e., $x \cdot y$ rather than $x * y$)

and is often omitted altogether. We also simplify the notation by referring to the group, when the binary operation is assumed known, as G,
rather than $(G, *)$.

EXAMPLES:

a. $(\mathbb{Z}, +)$, the integers with standard addition.

b. $(\mathbb{R} \setminus \{0\}, \cdot)$, the nonzero real numbers, with standard multiplication.

c. $S_n = (S_n, \cdot)$, the *symmetric group on* $[1, \dots, n]$; n a positive integer.
The elements of S_n are all the permutations σ of the set $[1, \dots, n]$,
i.e., the set of the bijections (1-1 maps) of $[1, \dots, n]$ onto itself.

The group operation is *composition*: for $\sigma, \tau \in S_n$ we define $\tau \cdot \sigma$
by: $(\tau \cdot \sigma)(j) = \tau(\sigma(j))$ for all j in $[1, n]$.

The first two examples are commutative; the third is not if $n > 2$.

1.1.2   Fields.

DEFINITION: A *field* $(\mathbb{F}, +, \cdot)$ is a set $\mathbb{F}$ endowed with two binary
operations, *addition*, $(a, b) \mapsto a + b$, and *multiplication*, $(a, b) \mapsto a \cdot b$,
(usually written simply as ab), such that:

F-1 $(\mathbb{F}, +)$ is a commutative group, whose identity element is denoted
by 0.

F-2 $(\mathbb{F} \setminus \{0\}, \cdot)$ is a commutative group, whose identity element is
denoted by 1. It is *the multiplicative group of* $\mathbb{F}$.

F-3   Addition and multiplication are related by the *distributive law*:

$$a(b + c) = ab + ac.$$

EXAMPLES:

a. $\mathbb{Q}$, the field of rational numbers.

b. $\mathbb{R}$, the field of real numbers.

c. $\mathbb{C}$, the field of complex numbers.

d. $\mathbb{Z}_2$ denotes the field consisting of the two elements 0, 1, with addition and multiplication defined mod 2 (so that $1 + 1 = 0$).

More generally, if p is a prime, the set $\mathbb{Z}_p$ of residue classes mod p, with addition and multiplication mod p, is a field. (See exercise **ex1.1.4**.)

The fields $\mathbb{Q}$, $\mathbb{R}$, and $\mathbb{C}$ are familiar, and are the most commonly used. Less familiar, yet very useful, are the finite fields $\mathbb{Z}_p$, mentioned above. Also important are *extensions* of a given field, see A.5.5.

EXERCISES FOR SECTION 1.1

ex1.1.1 Verify that S_n is not commutative if $n > 2$.

ex1.1.2 Show that if $\mathbb{F}$ is a field, then $0 \cdot a = 0$ for all $a \in \mathbb{F}$, and if $ab = 0$, then $a = 0$ or $b = 0$.

ex1.1.3 Verify that $\mathbb{Z}_3 = \{0, 1, 2\}$ is a field if addition and multiplication are defined mod 3, i.e., we add and multiply as usual, and if the result is ≥ 3, subtract 3; thus $1 + 1 = 2$ but $2 \times 2 = 1$.

Why is $\mathbb{Z}_4$, defined similarly—as the set $\{0, 1, 2, 3\}$ with addition and multiplication defined mod 4—not a field?

ex1.1.4 Let $p > 1$ be a positive integer. Recall that two integers m, n are *congruent mod* p, written $n \equiv m \pmod{p}$, if $n - m$ is divisible by p. This is an equivalence relation (see appendix A.1). For $m \in \mathbb{Z}$, denote by $\tilde{m}$ the coset (equivalence class) of m, that is, the set of all integers n such that $n \equiv m \pmod{p}$.

a. Every integer is congruent mod p to one of the numbers $[0, 1, \ldots, p-1]$. In other words, there is a 1-1 correspondence between $\mathbb{Z}_p$, the set of cosets (mod p), and the integers $[0, 1, \ldots, p-1]$.

b. As in subsection 1.2.4 above, we define the *quotient ring* $\mathbb{Z}_p = \mathbb{Z}/p\mathbb{Z}$ (both notations are common) as the space whose elements are the cosets (mod p) in $\mathbb{Z}$, and define addition and multiplication by: $\tilde{m} + \tilde{n} = \widetilde{(m+n)}$ and $\tilde{m} \cdot \tilde{n} = \widetilde{m \cdot n}$. Prove that the addition and multiplication so defined are associative, commutative, and satisfy the distributive law.

c. Prove that $\mathbb{Z}_p$, endowed with these operations, is a field if and only if p is prime.

Hint: You may use the following fact: if p is a prime, and both n and m are *not* divisible by p, then nm is not divisible by p. Show that this implies that if $\tilde{n} \neq 0$ in $\mathbb{Z}_p$, then $\{\tilde{n}\tilde{m} : \tilde{m} \in \mathbb{Z}_p\}$ covers all of $\mathbb{Z}_p$.

1.2   Vector spaces

1.2.1 DEFINITION. *A vector space $\mathscr{V}$ over a field* $\mathbb{F}$ is an abelian group $(\mathscr{V}, +)$, for which a binary product, $(a, v) \mapsto av$, of $\mathbb{F} \times \mathscr{V}$ into $\mathscr{V}$ is defined, satisfying the following axioms for all $a, b \in \mathbb{F}$ and $u, v \in \mathscr{V}$:

VS 1. $1v = v$,

VS 2. $(ab)v = a(bv)$,

VS 3. $(a + b)v = av + bv$ and $a(v + u) = av + au$.

Other familiar properties may be derived from these. For example, for every $v \in \mathscr{V}$, $0v = (1 - 1)v = v - v = 0$.

The elements of $\mathscr{V}$ are usually referred to as *vectors*; the elements of the underlying field as *scalars*.

Observe that in the equality $(ab)v = a(bv)$ the multiplication (ab) is within $\mathbb{F}$; the others are products of a vector by a scalar. In $(a+b)v = av + bv$, the addition on the left is the addition in $\mathbb{F}$, while that on the right is the addition in $\mathscr{V}$.

Most of the notions and results we discuss are valid for vector spaces over arbitrary fields. When the underlying field does not need to be specified, we denote it by the generic $\mathbb{F}$.

Results that apply to vector spaces over specific fields, or to vector spaces over fields satisfying some additional conditions, will be stated explicitly in terms of the appropriate field or the additional conditions.

If the underlying field is $\mathbb{R}$ or $\mathbb{C}$, then the vector space is called a *real vector space* or a *complex vector space*, respectively.

Vector spaces may also have additional structures: geometric, such as *inner-product*, which we study in Chapter 6; or algebraic, such as multiplication, as discussed in A.5.6.

EXAMPLES: The following sets are all vector spaces over the indicated fields.

a. $\mathbb{F}^n$, the space of all $\mathbb{F}$-valued n-tuples $[a_1, \ldots, a_n]$ with addition and multiplication by scalars defined by

$$[a_1, \ldots, a_n] + [b_1, \ldots, b_n] = [a_1 + b_1, \ldots, a_n + b_n],$$
$$c[a_1, \ldots, a_n] = [ca_1, \ldots, ca_n].$$

We may write the n-tuples as rows, as we did here, or as columns. If we want to specify that the vectors are written as columns or as rows, we write $\mathbb{F}^n_c$ or $\mathbb{F}^n_r$, respectively.

b. $\mathscr{M}(n, m; \mathbb{F})$, the space of all $\mathbb{F}$-valued $n \times m$ matrices; that is, arrays

$$A = \begin{bmatrix} a_{11} & \cdots & a_{1m} \\ a_{21} & \cdots & a_{2m} \\ \vdots & \cdots & \vdots \\ a_{n1} & \cdots & a_{nm} \end{bmatrix}$$

with entries from $\mathbb{F}$. We sometimes write $A = [a_{ij}]$ when the dimensions of the matrix are assumed known, to save space.

The addition and multiplication by scalars are again done entry by entry. As a vector space, $\mathscr{M}(n, m; \mathbb{F})$ is virtually identical with $\mathbb{F}^{mn}$.

We write $\mathscr{M}(n; \mathbb{F})$ instead of $\mathscr{M}(n, n; \mathbb{F})$, and if the underlying field is either known implicitly, or assumed explicitly, then we often write simply $\mathscr{M}(n, m)$ or $\mathscr{M}(n)$, as the case may be.

c. $\mathbb{F}[x]$, the space[1] of all polynomials $\sum a_n x^n$ with coefficients from $\mathbb{F}$. Addition and multiplication by scalars are defined formally either as the standard addition and multiplication of functions (of the variable x), or by adding (and multiplying by scalars) the corresponding coefficients. The two ways define the same operations.

More generally, the set $\mathbb{F}[x_1, \ldots, x_k]$ of all polynomials in k variables over $\mathbb{F}$ is a vector space (in fact—an algebra).

---

[1]$\mathbb{F}[x]$ is an algebra over $\mathbb{F}$, i.e., a vector space with the additional operation of multiplication. See A.5.6.

d. Let X be a finite set. $C(X)$ denotes the set of all complex-valued functions on X, with the standard addition of functions, and multiplication of functions by scalars.

e. The set $C_{\mathbb{R}}([0,1])$ of all continuous real-valued functions f on $[0,1]$, and the set $C([0,1])$ of all continuous complex-valued functions f on $[0,1]$, with the standard operations of addition of functions and of multiplication of functions by scalars. $C_{\mathbb{R}}([0,1])$ is a real vector space. $C([0,1])$ is naturally a complex vector space, but becomes a real vector space if we limit the allowable scalars to real numbers only.

f. The set $C^{\infty}([-1,1])$ of all infinitely differentiable real-valued functions f on $[-1,1]$, with the standard operations on functions is a real vector space.

g. The set $\mathscr{T}_N$ of 2π-periodic trigonometric polynomials of degree bounded by N; that is, the functions of the form $\sum_{|n|\leq N} a_n e^{inx}$. The underlying field may be $\mathbb{C}$ or $\mathbb{R}$, and the operations are the standard addition of functions and multiplication of functions by scalars.

Similarly, the space $\mathscr{T}_{N,M}$ of trigonometric polynomials in two variables of the form $\sum_{|n|\leq N,\ |m|\leq M} e^{i(nx+my)}$, with the same operations and the same underlying field(s).

h. The set of complex-valued functions f on $\mathbb{R}$ that satisfy the differential equation

$$3f'''(x) - \sin x f''(x) + 2f(x) = 0,$$

with the standard operations on functions. If we are interested in real-valued functions only, the underlying field is naturally $\mathbb{R}$. If we allow complex-valued functions we may choose either $\mathbb{C}$ or $\mathbb{R}$ as the underlying field.

1.2.2   Isomorphism. The expression "virtually identical" in the comparison above of $\mathscr{M}(n,m;\mathbb{F})$ with $\mathbb{F}^{mn}$ is not a proper mathematical term. The proper term here is *isomorphic*.

Recall (or see section A.2 in the appendix) that a map $\varphi: X \mapsto Y$ from a set X to a set Y is *bijective* if for every $y \in Y$ there is precisely one $x \in X$ such that $y = \varphi(x)$. Bijective maps are *invertible*—the inverse map is defined by:

$$(1.2.1) \qquad \varphi^{-1}(y) = x \quad \text{if} \quad y = \varphi(x).$$

Let $\mathscr{V}$ and $\mathscr{W}$ be vector spaces (over the same field).

DEFINITION: A map $\varphi: \mathscr{V} \mapsto \mathscr{W}$ is *linear* if for all scalars a, b and vectors $v_1, v_2 \in \mathscr{V}$

$$(1.2.2) \qquad \varphi(av_1 + bv_2) = a\varphi(v_1) + b\varphi(v_2).$$

A map $\varphi: \mathscr{V} \mapsto \mathscr{W}$ is *an isomorphism* if it is *both bijective and linear*. An *automorphism* is an isomorphism of a vector space $\mathscr{V}$ onto itself.

If φ is an isomorphism of $\mathscr{V}$ onto $\mathscr{W}$, then φ^{-1} is an isomorphism of $\mathscr{W}$ onto $\mathscr{V}$. This can be seen as follows: by (1.2.2),

$$(1.2.3) \qquad \begin{aligned} \varphi^{-1}(a_1\varphi v_1 + a_2\varphi v_2) &= \varphi^{-1}(\varphi(a_1 v_1 + a_2 v_2)) = a_1 v_1 + a_2 v_2 \\ &= a_1\varphi^{-1}(\varphi v_1) + a_2\varphi^{-1}(\varphi v_2), \end{aligned}$$

and, as φ is surjective, every vector in $\mathscr{W}$ is equal to φv for some $v \in \mathscr{V}$.

We say that $\mathscr{V}$ *and* $\mathscr{W}$ *are isomorphic* if there is an isomorphism φ of $\mathscr{V}$ onto $\mathscr{W}$, and the fact that the inverse φ^{-1} is also an isomorphism guarantees that the relation of being isomorphic is symmetric.

The identity map $\varphi(x) = x$ shows that the relation is reflexive and, since the composition of isomorphisms is an isomorphism, see exercise **ex1.2.2**, the relation is also transitive. In other words, the relations of being isomorphic is *an equivalence relation* (see A.1.2).

1.2.3 Subspaces. A (vector) *subspace* of a vector space $\mathscr{V}$ is a subset that is closed under the operations of addition and multiplication by scalars inherited from $\mathscr{V}$. In other words, $\mathscr{W} \subset \mathscr{V}$ is a subspace if for all scalars a_j and vectors $w_j \in \mathscr{W}$, $j = 1, 2$, the vectors $a_1 w_1 + a_2 w_2$ are in $\mathscr{W}$.

EXAMPLES:

a. *Solution-set of a system of homogeneous linear equations.*

Here $\mathscr{V} = \mathbb{F}^n$. Given the scalars a_{ij}, $1 \le i \le k$, $1 \le j \le n$, we consider the solution-set of the system of k homogeneous linear equations

$$(1.2.4) \qquad \sum_{j=1}^{n} a_{ij} x_j = 0, \quad i = 1, \ldots, k.$$

This is the set of all n-tuples, $[x_1, \ldots, x_n] \in \mathbb{F}^n$ (thought of as vectors), for which all k equations are satisfied. If both $[x_1, \ldots, x_n]$ and $[y_1, \ldots, y_n]$ are solutions of (1.2.4), and a and b are scalars, then for each i,

$$\sum_{j=1}^{n} a_{ij}(ax_j + by_j) = a \sum_{j=1}^{n} a_{ij} x_j + b \sum_{j=1}^{n} a_{ij} y_j = 0.$$

It follows that the solution-set of (1.2.4) is a subspace of $\mathbb{F}^n$.

b. In $\mathbb{F}[x]$, the space $\mathbb{F}_N[x]$ of polynomials $\sum_0^N a_n x^n$ of degree $\le N$. While $\mathbb{F}[x]$ is an algebra, $\mathbb{F}_N[x]$ is not an algebra; why?

c. In the space $C^\infty(\mathbb{R})$ of all infinitely differentiable, real-valued functions f on $\mathbb{R}$ with the standard operations, the set of functions f that satisfy the differential equation

$$f'''(x) - 5f''(x) + 2f'(x) - f(x) = 0.$$

If we consider complex-valued solutions, then the field of scalars can be $\mathbb{R}$ or $\mathbb{C}$.

d. Subspaces of $\mathscr{M}(n)$:

The set of *diagonal matrices*—the $n \times n$ matrices with zero entries off the main diagonal, i.e., $a_{ij} = 0$ for $i \ne j$.

The set of *symmetric matrices*—the $n \times n$ matrices whose entries satisfy $a_{ij} = a_{ji}$.

The set of *skew-symmetric matrices*—the $n \times n$ matrices whose entries satisfy $a_{ij} = -a_{ji}$.

The set of *lower triangular matrices*—the $n \times n$ matrices with zero entries above the main diagonal, i.e., $a_{ij} = 0$ for $i < j$.

Similarly, the set of *upper triangular matrices*, i.e., those for which $a_{ij} = 0$ for $i > j$.

Remark: In view of the isomorphism between $\mathcal{M}(n)$ and $\mathbb{F}^{n^2}$, these subspaces can be viewed as special cases of example *a*.

e. *Intersection of subspaces:* If $\mathcal{W}_j$ are subspaces of a space $\mathcal{V}$, $j \in J$, (the index set J can be finite or infinite), then $\bigcap \mathcal{W}_j$ is a subspace of $\mathcal{V}$.

f. *The sum of subspaces:* If $\mathcal{W}_j$, $j \in J$, are subspaces of $\mathcal{V}$, their sum is the set:

$$\sum \mathcal{W}_j = \bigcup_{J_1 \subset J} \{v : v = \sum_{j \in J_1} v_j, \ v_j \in \mathcal{W}_j\},$$

where the union extends over the collection of *finite* subsets of J.

Don't confuse the *sum of subspaces* with *the union of subspaces*, which is seldom a subspace, see exercises **ex1.2.5** and **ex1.2.4**.

g. *The span of a subset:* Given $E \subset \mathcal{V}$, a *linear combination* of elements of E is *a finite sum* of the form $\sum a_j v_j$, $a_j \in \mathbb{F}$, $v_j \in E$. The set of all the linear combinations of elements of E is a subspace of $\mathcal{V}$, called *the span* of E and denoted by $\mathrm{span}[E]$.

1.2.4 Quotient spaces. A subspace $\mathcal{W}$ of a vector space $\mathcal{V}$ defines an equivalence relation (see the appendix, section A.1) in $\mathcal{V}$:

(1.2.5) $\qquad x \equiv y \pmod{\mathcal{W}} \qquad$ if $\quad x - y \in \mathcal{W}$.

In order to establish that this is indeed an equivalence relation, we need to verify that it is reflexive, symmetric, and transitive:

a. *reflexive:* $x \equiv x$, since $x - x = 0 \in \mathcal{W}$,

b. *symmetric:* $x \equiv y \iff y \equiv x$, since $x - y \in \mathscr{W}$ if and only if
$y - x = -(x - y) \in \mathscr{W}$,

c. *transitive:* If both $x \equiv y$ and $y \equiv z$, then $x \equiv z$. This follows from: if
both $(x - y)$ and $(y - z)$ are in $\mathscr{W}$, then so is $x - z = (x - y) + (y - z)$.

This equivalence relation partitions $\mathscr{V}$ into equivalence classes,
called the *cosets* of $\mathscr{W}$ in $\mathscr{V}$. For $x \in \mathscr{V}$, the coset of $\mathscr{W}$ that contains x is the set $\tilde{x} = x + \mathscr{W} = \{v = x + w : w \in \mathscr{W}\}$—the "translate"
of $\mathscr{W}$ by x.

We define the *quotient space* $\mathscr{V}/\mathscr{W}$ to be the space whose elements
are the cosets of $\mathscr{W}$ in $\mathscr{V}$, and the operations of addition and multiplication by scalars are defined as follows. If $\tilde{x} = x + \mathscr{W}$ and $\tilde{y} = y + \mathscr{W}$
are cosets, and $a \in \mathbb{F}$, then

(1.2.6) $\tilde{x} + \tilde{y} = x + y + \mathscr{W} = \widetilde{x + y}$ and $a\tilde{x} = \widetilde{ax}$.

The definition needs justification. We defined the sum of two
cosets by taking a representative element from each, and taking the
coset that contains their sum as the sum of the cosets. We need to show
that the result is well defined, i.e., that it does not depend on the choice
of the representatives in the cosets. In other words, we need to verify
that if $x \equiv x_1$ (mod $\mathscr{W}$) and $y \equiv y_1$ (mod $\mathscr{W}$), then $x + y \equiv x_1 + y_1$
(mod $\mathscr{W}$). Now, if $x = x_1 + w$, $y = y_1 + w'$ with $w, w' \in \mathscr{W}$, then
$x + y = x_1 + w + y_1 + w' = x_1 + y_1 + w + w'$, and since $w + w' \in \mathscr{W}$,
we have $x + y \equiv x_1 + y_1$ (mod $\mathscr{W}$).

The definition of $a\tilde{x}$ is justified similarly: if $x \equiv x_1$ (mod $\mathscr{W}$), then
$ax - ax_1 = a(x - x_1)$, and since $\mathscr{W}$ is a *subspace*, it is closed under
multiplication by scalars, $a(x - x_1) \in \mathscr{W}$, and $ax \equiv ax_1$ (mod $\mathscr{W}$) .

1.2.5 Direct sums. If $\mathscr{V}_1, \ldots, \mathscr{V}_k$ are vector spaces over $\mathbb{F}$, *the (formal)
direct sum*

$$\bigoplus_1^k \mathscr{V}_j = \mathscr{V}_1 \oplus \cdots \oplus \mathscr{V}_k$$

is the set $\{(v_1, \ldots, v_k) : v_j \in \mathcal{V}_j\}$ in which addition and multiplication by scalars are defined by:

$$(v_1, \ldots, v_k) + (u_1, \ldots, u_k) = (v_1 + u_1, \ldots, v_k + u_k),$$
$$a(v_1, \ldots, v_k) = (av_1, \ldots, av_k).$$

The direct sum of vector spaces is clearly a vector space.

In the case that $\mathcal{V}_1, \ldots, \mathcal{V}_k$ are all subspaces of the same vector space $\mathcal{V}$, we defined their *sum*, $\sum \mathcal{V}_j = \{v : v = \sum_{j=1}^k v_j, \ v_j \in \mathcal{V}_j\}$, in example f of 1.2.3.

The "natural" map of $\mathcal{V}_1 \oplus \cdots \oplus \mathcal{V}_k$ into $\mathcal{V}_1 + \cdots + \mathcal{V}_k$, defined by

(1.2.7) $$\Phi((v_1, \ldots, v_k)) = \sum_1^k v_j,$$

is clearly linear and surjective. It is an isomorphism when the subspaces are *independent*.

DEFINITION: The subspaces $\mathcal{V}_j$, $j = 1, \ldots, k$, of a vector space $\mathcal{V}$ are *independent* if $\sum v_j = 0$ with $v_j \in \mathcal{V}_j$ implies that $v_j = 0$ for all j.[2]

Proposition. *Let $\mathcal{V}_j$ be subspaces of $\mathcal{V}$. The map Φ defined by (1.2.7) is an isomorphism if and only if the subspaces are independent.*

PROOF: Φ is clearly linear and surjective. It is injective if and only if every vector in the range has a unique preimage, that is, if

(1.2.8) $$v_j', v_j'' \in \mathcal{V}_j \qquad \text{and} \qquad v_1'' + \cdots + v_k'' = v_1' + \cdots + v_k'$$

implies that $v_j'' = v_j'$ for every j. Subtracting and writing $v_j = v_j'' - v_j'$, (1.2.8) is equivalent to: $\sum v_j = 0$ with $v_j \in \mathcal{V}_j$. The subspaces are independent if and only if this implies that $v_j = 0$ for all j. ◀

In view of the proposition, we refer to the sum $\sum \mathcal{W}_j$ of *independent* subspaces of a vector space as their *direct sum*, and write $\bigoplus \mathcal{W}_j$ instead of $\sum \mathcal{W}_j$.

If $\mathcal{V} = \mathcal{U} \oplus \mathcal{W}$, we refer to $\mathcal{U}$ as *a complement* of $\mathcal{W}$ in $\mathcal{V}$, and vice versa.

---

[2]Properly speaking: the set $\{\mathcal{V}_j\}$ is independent.

1.2.6   Tensor products. Let $\mathcal{V}$ and $\mathcal{U}$ be vector spaces over $\mathbb{F}$. Let $\widetilde{\mathcal{V} \otimes \mathcal{U}}$ be the set of all the (finite) formal sums $\sum a_j v_j \otimes u_j$, where $a_j \in \mathbb{F}$, $v_j \in \mathcal{V}$ and $u_j \in \mathcal{U}$. We define addition formally by

$$\sum_{j \in J_1} a_j v_j \otimes u_j + \sum_{j \in J_2} a_j v_j \otimes u_j = \sum_{j \in J_1 \cup J_2} a_j v_j \otimes u_j,$$

and we define multiplication by scalars by

$$a \sum a_j v_j \otimes u_j = \sum (a a_j) v_j \otimes u_j.$$

With these definitions, $\widetilde{\mathcal{V} \otimes \mathcal{U}}$ is a vector space over $\mathbb{F}$.

The *tensor product* $\mathcal{V} \otimes \mathcal{U}$ is, by definition, the quotient of $\widetilde{\mathcal{V} \otimes \mathcal{U}}$ by the subspace $[\widetilde{\mathcal{V} \otimes \mathcal{U}}]_0$ spanned by the elements of the form

$$\begin{array}{cl} \textbf{a.} & (v_1 + v_2) \otimes u - (v_1 \otimes u + v_2 \otimes u), \\ (1.2.9) \quad \textbf{b.} & v \otimes (u_1 + u_2) - (v \otimes u_1 + v \otimes u_2), \\ \textbf{c.} & a(v \otimes u) - (av) \otimes u, \quad (av) \otimes u - v \otimes (au), \end{array}$$

for all $v, v_j \in \mathcal{V}$, $u, u_j \in \mathcal{U}$, and $a \in \mathbb{F}$.

In other words, $\mathcal{V} \otimes \mathcal{U}$ is the space of formal sums $\sum a_j v_j \otimes u_j$ *modulo the equivalence relation* generated by:

$$\begin{array}{cl} \textbf{a.} & (v_1 + v_2) \otimes u \equiv v_1 \otimes u + v_2 \otimes u, \\ (1.2.10) \quad \textbf{b.} & v \otimes (u_1 + u_2) \equiv v \otimes u_1 + v \otimes u_2, \\ \textbf{c.} & a(v \otimes u) \equiv (av) \otimes u \equiv v \otimes (au). \end{array}$$

EXAMPLE: If $\mathcal{V} = \mathbb{F}[x]$ and $\mathcal{U} = \mathbb{F}[y]$, and we define the map Φ of $\widetilde{\mathcal{V} \otimes \mathcal{U}}$ into $\mathbb{F}[x, y]$ by:

$$(1.2.11) \qquad \Phi \colon \sum_j a_j p_j(x) \otimes q_j(y) \mapsto \sum_j a_j p_j(x) q_j(y) \in \mathbb{F}[x, y],$$

then all the elements of $[\widetilde{\mathcal{V} \otimes \mathcal{U}}]_0$ are mapped to zero, so that all the elements in an equivalence class modulo $[\widetilde{\mathcal{V} \otimes \mathcal{U}}]_0$ are mapped to the same polynomial. For example, every formal sum in $\widetilde{\mathcal{V} \otimes \mathcal{U}}$ that is equivalent to $p(x) \otimes q(y)$ is mapped to $p(x)q(y)$. It follows that Φ induces a map of the quotient space $\mathcal{V} \otimes \mathcal{U}$ onto $\mathbb{F}[x,y]$

$$(1.2.12) \qquad \overline{\Phi}: \sum_j p_j(x) \otimes q_j(y) \mapsto \sum_j p_j(x)q_j(y).$$

$\overline{\Phi}$ is an isomorphism, and the spaces $\mathcal{V} \otimes \mathcal{U}$ and $\mathbb{F}[x,y]$ are isomorphic.

EXERCISES FOR SECTION 1.2

ex1.2.1 Verify that $\mathbb{R}$ is a vector space over $\mathbb{Q}$, and that $\mathbb{C}$ is a vector space over either $\mathbb{Q}$ or $\mathbb{R}$.

ex1.2.2 Let $\mathcal{V}_j, j = 1,2,3$, be vector spaces over the same field $\mathbb{F}$. Let $\varphi_1 : \mathcal{V}_1 \mapsto \mathcal{V}_2$ and $\varphi_2 : \mathcal{V}_2 \mapsto \mathcal{V}_3$ be isomorphisms. Prove that $\varphi_2\varphi_1$ is an isomorphism of $\mathcal{V}_1$ onto $\mathcal{V}_3$. Conclude that *isomorphism* is an equivalence relation for vector spaces (defined over the same field $\mathbb{F}$).

ex1.2.3 Verify that the intersection of subspaces is a subspace.

ex1.2.4 Let $\mathcal{U}$ and $\mathcal{W}$ be subspaces of a vector space $\mathcal{V}$, and neither of them contains the other. Show that $\mathcal{U} \cup \mathcal{W}$ is *not* a subspace.
Hint: Take $u \in \mathcal{U} \setminus \mathcal{W}$, $w \in \mathcal{W} \setminus \mathcal{U}$ and consider $u + w$.

ex1.2.5 Verify that the sum of subspaces is a subspace, and prove that

$$\sum \mathcal{W}_j = \mathsf{span}[\bigcup \mathcal{W}_j].$$

ex1.2.6 Check that for every $E \subset \mathcal{V}$, $\mathsf{span}[E]$ is a subspace of $\mathcal{V}$, and is contained in every subspace that contains E.

ex1.2.7 If $\mathcal{V}_1$ is a subspace of $\mathcal{V}$ and φ is an isomorphism of $\mathcal{V}$ onto $\mathcal{W}$, then $\varphi\mathcal{V}_1$ is a subspace of $\mathcal{W}$.

ex1.2.8 Describe all the complements in $\mathbb{R}^2$ of the subspace $X = \{(x,0) : x \in \mathbb{R}\}$.

ex1.2.9 Prove that two subspaces $\mathcal{V}_1$ and $\mathcal{V}_2$ in a vector space are independent if $\mathcal{V}_1 \cap \mathcal{V}_2 = \{0\}$.

ex1.2.10 Prove that the subspaces $\mathcal{W}_j \subset \mathcal{V}$, $j = 1,\ldots,N$ are independent if and only if $\mathcal{W}_j \cap \sum_{l \neq j} \mathcal{W}_l = \{0\}$ for all j.

ex1.2.11 Let $\mathcal{U}$ and $\mathcal{W}$ be subspaces of a vector space $\mathcal{V}$; then $\mathcal{U}$ is a subspace of $\mathcal{U} + \mathcal{W}$, and $\mathcal{U} \cap \mathcal{W}$ is a subspace of $\mathcal{W}$. Prove that the quotient spaces

$$(1.2.13) \qquad (\mathcal{U} + \mathcal{W})/\mathcal{U} \quad \text{and} \quad \mathcal{W}/(\mathcal{U} \cap \mathcal{W})$$

are isomorphic.

Hint: A coset of $\mathcal{U}$ in $(\mathcal{U} + \mathcal{W})$ has the form $w + \mathcal{U}$, and $w_1 + \mathcal{U} = w_2 + \mathcal{U}$ if and only if $w_1 - w_2 \in \mathcal{U}$.

$*$**ex1.2.12** Assuming that $\mathbb{F}$ is infinite,[3] and $\mathcal{V}$ is a vector space over $\mathbb{F}$, show that the union of a finite number of subspaces of $\mathcal{V}$, none of which contains all the others, is not a subspace.

Hint: Let $\mathcal{V}_j$, $j = 1, \dots, k$ be the subspaces in question. Show that there is no loss in generality in assuming that their union spans $\mathcal{V}$. Now you need to show that $\bigcup \mathcal{V}_j$ is not all of $\mathcal{V}$. Show that there is no loss of generality in assuming that $\mathcal{V}_1$ is not contained in the union of the others. Now take $v_1 \in \mathcal{V}_1 \setminus \bigcup_{j \neq 1} \mathcal{V}_j$, and $w \notin \mathcal{V}_1$; show that $av_1 + w \in \bigcup \mathcal{V}_j$, $a \in \mathbb{F}$, for no more than k values of a.

1.3   Linear dependence, bases, and dimension

Let $\mathcal{V}$ be a vector space. A *linear combination of vectors* $v_1, \dots, v_k$ is a sum of the form $v = \sum a_j v_j$ with scalar coefficients a_j.

A linear combination is *nontrivial* if at least one of the coefficients is not zero; otherwise it is *trivial*.

1.3.1 Recall that if A is a subset of a vector space $\mathcal{V}$, then $\text{span}[A]$, the set of all linear combinations of elements in A, is a subspace of $\mathcal{V}$, (see example f of 1.2.3).

DEFINITION: The set $A \subset \mathcal{V}$ is a *spanning set* if $\text{span}[A] = \mathcal{V}$.

1.3.2 DEFINITION. The set $A \subset \mathcal{V}$ is *linearly independent* if for every sequence $\{v_1, \dots, v_l\}$ of distinct vectors in A, the only vanishing linear combination of the v_j's is trivial; that is, if $\sum a_j v_j = 0$, then $a_j = 0$ for all j.[4]

---

[3]If $\mathbb{F}$ is finite, then $\mathbb{F}^n = \cup_{v \in \mathbb{F}^n} \text{span}[v]$ is a finite union of subspaces.

[4]Independence is a property of the set A; however, we often say, by abuse of language, that *the vectors (in A) are independent*.

If the set A is finite, we enumerate its elements as $v_1, \ldots, v_m$ and write the elements in its span as $\sum a_j v_j$. By definition, independence of A means that the representation of $v = 0$ as a linear combination of elements from A is unique. Notice, however, that this implies that the representation of *every* vector in span$[A]$ is unique. In fact, the equality $\sum_1^l a_j v_j = \sum_1^l b_j v_j$ implies $\sum_1^l (a_j - b_j) v_j = 0$ so that $a_j = b_j$ for all j.

A vector v is *linearly dependent* on a set A if it can be represented as a linear combination of vectors from A, that is, if $v \in \text{span}[A]$.

1.3.3 A *minimal spanning set* is a spanning set such that no proper subset thereof is spanning.

A *maximal independent set* is an independent set such that no set that contains it properly is independent.

Lemma.

a. *A minimal spanning set is independent.*

b. *A maximal independent set is spanning.*

PROOF: **a.** Let A be a minimal spanning set. If $\sum a_j v_j = 0$, with distinct $v_j \in A$, and for some k, $a_k \neq 0$, then $v_k = -a_k^{-1} \sum_{j \neq k} a_j v_j$. This permits the substitution of v_k in any linear combination by the combination of the other v_j's, and shows that v_k is redundant: the span of $\{v_j : j \neq k\}$ is the same as the original span, contradicting the minimality assumption.

b. If B is independent and u is not in span$[B]$, then the union $\{u\} \cup B$ is independent: otherwise there would exist $\{v_1, \ldots, v_l\} \subset B$ and coefficients d and c_j, not all zero, such that $du + \sum c_j v_j = 0$.

If $d \neq 0$, then $u = -d^{-1} \sum c_j v_j$ and u would be in span$[v_1, \ldots, v_l] \subset$ span$[B]$, contradicting the assumption $u \notin \text{span}[B]$.

If $d = 0$, we have $\sum c_j v_j = 0$ with some nonvanishing coefficients, contradicting the assumption that B is independent. It follows that if B is maximal independent, then $u \in \text{span}[B]$ for every $u \in \mathcal{V}$; in other words: B is spanning. ◄

DEFINITION: A *basis for* $\mathcal{V}$ is a set $B \subset \mathcal{V}$ which is both spanning and independent. A vector space is *finite-dimensional* if it has a finite basis.

Thus, $\{v_1, \ldots, v_n\}$ is a basis for $\mathscr{V}$ if and only if every $v \in \mathscr{V}$ has a unique representation as a linear combination of $\{v_1, \ldots, v_n\}$. This representation, $v = \sum a_j v_j$, is the *expansion of v relative to the basis* $\{v_1, \ldots, v_n\}$.

By the lemma, a minimal spanning set is a basis, and a maximal independent set is a basis.

1.3.4 Proposition. *If $\mathscr{V}$ is finite-dimensional, then:*

a. *Every finite spanning set can be trimmed to a basis.*

b. *Every independent set can be expanded to a basis.*

PROOF: **a.** Let $\{v_j\}_{j=1}^N$ be a spanning set for $\mathscr{V}$. Call a vector v_l *inessential* if it is linearly dependent on $\{v_j\}_{j=1}^{l-1}$, and *essential* otherwise. *This classification clearly depends on the order in which the vectors are enumerated.* Observe that an inessential v_l is linearly dependent on the *essential* vectors preceding it.

Remove the inessential vectors. Since every v_j is either essential or linearly dependent on the preceding essential vectors, the essential vectors span $\mathscr{V}$ and are independent, hence they form a basis.

b. Let $\{u_j\}_{j=1}^k$ be independent, and let $\{e_j\}_{j=1}^n$ be a basis for $\mathscr{V}$. Write $w_j = u_j$ for $j = 1, \ldots, k$, and $w_{k+j} = e_j$ for $j = 1, \ldots, n$. The sequence $\{w_j\}$ contains the basis $\{e_j\}$ and is therefore spanning. Now remove, as in part **a**, the inessential vectors to obtain a basis, and observe that the first k vectors, namely $\{u_j\}_{j=1}^k$, are all essential, and hence form part of the basis.   ◀

Remarks: The statement of **a** and its proof are valid for infinite spanning sequences as well. See also 1.3.8.

The argument of **b** is refined somewhat in the following subsection.

EXAMPLES:

a. *The standard basis* for $\mathbb{F}^n$: we write e_j for the vector whose j'th entry is equal to 1 and all the other entries are zero. Then $\{e_1, \ldots, e_n\}$

is a basis for $\mathbb{F}^n$, and the unique representation of $v = \begin{bmatrix} a_1 \\ \vdots \\ a_n \end{bmatrix}$ in

terms of this basis is $v = \sum a_j e_j$.

b. *The standard basis* for $\mathscr{M}(n,m)$: let E_{ij} denote the $n \times m$ matrix whose ij'th entry is 1 and all the other entries are zero. Then $\{E_{ij}\}$ is a basis for $\mathscr{M}(n,m)$, and the expansion of $A = [a_{ij}] \in \mathscr{M}(n,m)$ is $A = \sum a_{ij}E_{ij}$.

c. The space $\mathbb{F}[x]$ does not have a finite basis. The infinite sequence $\{x^n\}_{n=0}^{\infty}$ is both linearly independent and spanning, that is, a basis. As we see in the following subsection, the existence of an infinite basis, even of an infinite independent set, precludes a finite basis and the space is *infinite-dimensional*.

Notice, however, that the subspace $\mathbb{F}_N[x]$, consisting of all the polynomials of degree at most N, is finite-dimensional since the set of $N+1$ vectors, $\{x^n\}_{n=0}^{N}$, is a basis.

1.3.5 Steinitz' lemma and the definition of dimension.

Lemma (Steinitz). *If* $\operatorname{span}[v_1, \ldots, v_n] = \mathscr{V}$ *and* $\{u_1, \ldots, u_m\}$ *is linearly independent in* $\mathscr{V}$*, then the vectors* v_j *can be (re)ordered so that, for every* $k = 1, \ldots, m$*, the sequence* $\{u_1, \ldots, u_k, v_{k+1}, \ldots, v_n\}$ *spans* $\mathscr{V}$*. In particular,* $m \leq n$*.*

PROOF: Write $u_1 = \sum a_j v_j$; this is possible since $\operatorname{span}[v_1, \ldots, v_n] = \mathscr{V}$. Reorder the v_j's, if necessary, to guarantee that $a_1 \neq 0$.

Now $v_1 = a_1^{-1}(u_1 - \sum_{j=2}^{n} a_j v_j)$, so that $\operatorname{span}[u_1, v_2, \ldots, v_n]$ contains every v_j and hence is equal to $\mathscr{V}$.

Continue inductively: assume that, having reordered v_j's, if necessary, we have $\{u_1, \ldots, u_k, v_{k+1}, \ldots, v_n\}$ spans $\mathscr{V}$.

If $k = m$ we are done. If $k < m$ write

(1.3.1)
$$u_{k+1} = \sum_{j=1}^{k} a_j u_j + \sum_{j=k+1}^{n} b_j v_j.$$

Since $\{u_1,\dots,u_m\}$ is linearly independent, at least one of the coefficients b_j is not zero. Reordering, if necessary, the remaining v_j's, we may assume that $b_{k+1} \neq 0$. Now rewrite (1.3.1):

$$(1.3.2) \qquad v_{k+1} = -b_{k+1}^{-1}\Big(\sum_{j=1}^{k} a_j u_j - u_{k+1} + \sum_{j=k+2}^{n} b_j v_j\Big),$$

so that $v_{k+1} \in \mathsf{span}[u_1,\dots,u_{k+1},v_{k+2},\dots,v_n]$, and, once again, the span is $\mathscr{V}$. Repeating the step m times proves the lemma.   ◀

Theorem. *If* $\{v_1,\dots,v_n\}$ *and* $\{u_1,\dots,u_m\}$ *are both bases, then* $m = n$.

PROOF: Since $\{v_1,\dots,v_n\}$ is spanning and $\{u_1,\dots,u_m\}$ is independent, we have $m \leq n$. Reversing the roles we have $n \leq m$.   ◀

Part of Steinitz' lemma repeats part **b** of proposition 1.3.4: in a finite-dimensional vector space, every independent set can be expanded to a basis by adding, if necessary, elements from any given spanning set. The additional information here, that any spanning set has at least as many elements as any independent set, is the basis for the current theorem, and enables the definition of *dimension*.

Recall that a vector space $\mathscr{V}$ is *finite-dimensional* if it has a finite basis.

DEFINITION: The *dimension* of a finite-dimensional vector space $\mathscr{V}$, denoted $\dim\mathscr{V}$, is the number of elements in any basis for $\mathscr{V}$. The definition is unambiguous, since all bases have the same cardinality.

As you are asked to check in exercise **ex1.3.6** below, a subspace $\mathscr{W}$ of a finite-dimensional space $\mathscr{V}$ is finite-dimensional and, unless $\mathscr{W} = \mathscr{V}$, the dimension $\dim\mathscr{W}$ of $\mathscr{W}$ is strictly lower than $\dim\mathscr{V}$.

The *codimension* of a subspace $\mathscr{W}$ in a finite-dimensional space $\mathscr{V}$ is, by definition, $\dim\mathscr{V} - \dim\mathscr{W}$.

1.3.6   Assume that $\mathscr{V}$ and $\mathscr{W}$ are finite-dimensional vector spaces over $\mathbb{F}$, and $\dim\mathscr{V} = \dim\mathscr{W} = d$. Let $\{v_j\}_{j=1}^{d}$ and $\{w_j\}_{j=1}^{d}$ be bases for $\mathscr{V}$ and $\mathscr{W}$ respectively.

Every element $v \in \mathcal{V}$ has a unique representation as a linear combination of the basis elements, so the map $\varphi \colon \mathcal{V} \mapsto \mathcal{W}$ defined by:

$$(1.3.3) \qquad \varphi\Big(\sum_1^d a_j v_j\Big) = \sum_1^d a_j w_j,$$

is unambiguous. Since $\{w_j\}$ is a basis of $\mathcal{W}$, the map φ is a bijection onto $\mathcal{W}$ and it clearly satisfies condition (1.2.2); in other words, *it is an isomorphism.* Conversely, if $\mathcal{V}$ is finite-dimensional and φ is an isomorphism of $\mathcal{V}$ onto a vector space $\mathcal{W}$, then the φ-image of a basis of $\mathcal{V}$ is a basis of $\mathcal{W}$. This proves the following theorem.

Theorem. *Two finite-dimensional vector spaces over the same field are isomorphic if and only if they have the same dimension.*

Remark: The definition (1.3.3) of the map φ can be viewed as a two-step process. The first step assigns to each basis element v_j its image w_j; the second "completes the definition by linearity": if φ is to be linear and $\varphi(v_j) = w_j$, then linearity forces, and in turn is guaranteed by, (1.3.3). The fact that $\{v_j\}$ is a basis guarantees that φ is well defined, and this independently of whether $\{w_j\}$ is a basis or not. Adding the assumption that $\{w_j\}$ is a basis for $\mathcal{W}$ guarantees that φ is bijective, and hence an isomorphism.

1.3.7 The following observation is sometimes useful.

Proposition. *Let $\mathcal{U}$ and $\mathcal{W}$ be subspaces of an n-dimensional space $\mathcal{V}$, and assume that* $\dim \mathcal{U} + \dim \mathcal{W} > n$. *Then* $\mathcal{U} \cap \mathcal{W} \neq \{0\}$.

PROOF: Let $\{u_j\}_{j=1}^l$ be a basis for $\mathcal{U}$ and $\{w_j\}_{j=1}^m$ be a basis for $\mathcal{W}$. Since $l + m > n$, the set $\{u_j\}_{j=1}^l \cup \{w_j\}_{j=1}^m$ is linearly dependent, i.e., there exists a nontrivial vanishing linear combination

$$\sum c_j u_j + \sum d_j w_j = 0.$$

If all the coefficients c_j were zero, we would have a vanishing nontrivial combination of the basis elements $\{w_j\}_{j=1}^m$, which is ruled out. Similarly not all the d_j's vanish. We now have in $\mathcal{U} \cap \mathcal{W}$ the nontrivial vector $v = \sum c_j u_j = -\sum d_j w_j$. ◀

1.3.8   Infinite-dimensional vector spaces. Many examples of vector spaces, in many areas of mathematics, are *infinite-dimensional,* i.e., they do not have finite bases. Assuming *the axiom of choice,* it can be shown that every spanning set in an arbitrary vector space $\mathcal{V}$ can be trimmed down to a *minimal spanning set,* i.e., one that does not contain a proper subset that is spanning. Likewise, every independent set is contained in a *maximal independent set,* i.e., one that is not properly contained in a larger linearly independent set.

A subset $\mathcal{B}$ which is either a maximal independent set or a minimal spanning set is a basis in the sense that every $v \in \mathcal{V}$ is equal to a unique *finite* linear combination of elements of $\mathcal{B}$. A basis, so defined, is called a *Hamel basis* for an infinite-dimensional space.

In a typical study involving infinite-dimensional vector spaces this is irrelevant! Infinite-dimensional spaces (of interest) usually come with a *topology,* which allows one to introduce the notion of *convergence.* In this context, *convergent infinite sums* are allowed, and bases are defined accordingly. The field devoted to the study of topological vector spaces is *Functional Analysis.*

EXERCISES FOR SECTION 1.3

ex1.3.1 Show that the set $\{v_j : 1 \le j \le k\}$ is linearly dependent if and only if $v_1 = 0$, or there exists $l \in [2, k]$ such that v_l is a linear combination of vectors in $\{v_j : 1 \le j \le l - 1\}$.

ex1.3.2 Let $\mathcal{V}$ be a vector space, and $\mathcal{W} \subset \mathcal{V}$ a subspace[5]. Let $v, u \in \mathcal{V} \setminus \mathcal{W}$, and assume that $u \in \text{span}[\mathcal{W}, v]$. Prove that $v \in \text{span}[\mathcal{W}, u]$.

ex1.3.3 What is the dimension of $\mathbb{C}^5$ considered as a vector space over $\mathbb{R}$?

ex1.3.4 Is $\mathbb{R}$ finite-dimensional over $\mathbb{Q}$?

ex1.3.5 Let $\mathcal{U}, \mathcal{W}$ be subspaces of a vector space $\mathcal{V}$, with $\mathcal{U} \cap \mathcal{W} = \{0\}$. Assume that $\{u_1, \dots, u_k\} \subset \mathcal{U}$ and $\{w_1, \dots, w_l\} \subset \mathcal{W}$ are linearly independent sets. Prove that $\{u_1, \dots, u_k\} \cup \{w_1, \dots, w_l\}$ is linearly independent.

ex1.3.6 Let $\mathcal{V}$ be finite-dimensional. Prove that every subspace $\mathcal{W} \subset \mathcal{V}$ is finite-dimensional, and that $\dim \mathcal{W} \le \dim \mathcal{V}$ with equality only if $\mathcal{W} = \mathcal{V}$.

---

[5] $\mathcal{V} \setminus \mathcal{W}$ denotes the difference set $\{v : v \in \mathcal{V} \text{ and } v \notin \mathcal{W}\}$.

ex1.3.7 Show that if $\mathscr{V}$ is finite-dimensional, then every subspace $\mathscr{W} \subset \mathscr{V}$ is a direct summand, i.e., there is a subspace $\mathscr{W}' \subset \mathscr{V}$ such that $\mathscr{V} = \mathscr{W} \oplus \mathscr{W}'$. Show by example that $\mathscr{W}'$ need not be unique (see exercise **ex1.2.8**).

ex1.3.8 Let $\mathscr{V}$ be a finite-dimensional vector space, and $\mathscr{A}$ a collection of subspaces of $\mathscr{V}$. Prove that there are one or more *minimal elements* in $\mathscr{A}$, that is $\mathscr{W} \in \mathscr{A}$ such that no element in $\mathscr{A}$ is a proper subspace of $\mathscr{W}$.

Similarly, show that $\mathscr{A}$ has one or more *maximal elements*, i.e., elements that are not contained in any other element of $\mathscr{A}$.

Remark: If $\mathscr{V}_j$ are subspaces of $\mathscr{V}$ such that none is contained in another, then every $\mathscr{V}_j$ is both minimal and maximal.

ex1.3.9 Assume that $\mathscr{V}$ is n-dimensional, and let $\mathscr{W}_j, j = 1, \ldots, k$ be subspaces of $\mathscr{V}$ such that, for $1 \leq j < k$, $\mathscr{W}_{j+1}$ is a proper subspace of $\mathscr{W}_j$. Prove that $k \leq n$.

ex1.3.10 Let $\mathscr{V}$ and $\mathscr{W}$ be finite-dimensional subspaces of a vector space. Prove that $\mathscr{V} + \mathscr{W}$ and $\mathscr{V} \cap \mathscr{W}$ are finite-dimensional and that

$$(1.3.4) \qquad \dim(\mathscr{V} \cap \mathscr{W}) + \dim(\mathscr{V} + \mathscr{W}) = \dim \mathscr{V} + \dim \mathscr{W}.$$

ex1.3.11 Repeat exercise **ex1.2.11** in the context of finite-dimensional spaces as follows:

a. Choose a basis $\{e_j\}$ for $\mathscr{U} \cap \mathscr{W}$;

b. Complete it to a basis for $\mathscr{U}$ by adding the vectors $\{u_k\}$;

c. Complete it also to a basis for $\mathscr{W}$ by adding the vectors $\{w_l\}$;

d. Check that $\{e_j\} \cup \{u_k\} \cup \{w_l\}$ is a basis for $\mathscr{U} + \mathscr{W}$;

e. Identify bases for the two quotient spaces invoved.

ex1.3.12 Show that if $\mathscr{W}_j$, $j = 1, \ldots, k$, are finite-dimensional subspaces of a vector space $\mathscr{V}$, then $\sum \mathscr{W}_j$ is finite-dimensional and $\dim \sum \mathscr{W}_j \leq \sum \dim \mathscr{W}_j$, with equality if and only if the subspaces $\mathscr{W}_j$ are independent.

ex1.3.13 Let $\mathscr{V}$ be an n-dimensional vector space, and let $\mathscr{V}_1 \subset \mathscr{V}$ be a subspace of dimension m.

a. Prove that the quotient space $\mathscr{V}/\mathscr{V}_1$ is finite-dimensional.

b. Let $\{v_1, \ldots, v_m\}$ be a basis for $\mathscr{V}_1$ and let $\{\tilde{w}_1, \ldots, \tilde{w}_k\}$ be a basis for $\mathscr{V}/\mathscr{V}_1$. For $j \in [1, k]$, let w_j be an element of the coset $\tilde{w}_j$.

Prove: $\{v_1, \ldots, v_m\} \cup \{w_1, \ldots, w_k\}$ is a basis for $\mathscr{V}$. Hence $k + m = n$.

ex1.3.14 Let $\mathcal{V}$ be a real vector space, and let $v_1, \ldots, v_p \in \mathcal{V}$ be linearly independent. Let $r_l = [a_{l,1}, \ldots, a_{l,p}]$, $1 \leq l \leq s$ be linearly independent vectors in $\mathbb{R}^p$. Prove that the vectors $u_l = \sum_1^p a_{l,j} v_j$, $l = 1, \ldots, s$, are linearly independent in $\mathcal{V}$.

ex1.3.15 Let $\mathcal{V}$ and $\mathcal{U}$ be finite-dimensional spaces over $\mathbb{F}$. Prove that the tensor product $\mathcal{V} \otimes \mathcal{U}$ is finite-dimensional. Specifically, if $\{e_j\}_{j=1}^n$ is a basis for $\mathcal{V}$, and $\{f_k\}_{k=1}^m$ is a basis for $\mathcal{U}$, then $\{e_j \otimes f_k\}$, $1 \leq j \leq n$, $1 \leq k \leq m$, is a basis for $\mathcal{V} \otimes \mathcal{U}$, so that $\dim \mathcal{V} \otimes \mathcal{U} = \dim \mathcal{V} \dim \mathcal{U}$.

∗**ex1.3.16** Assume that $\mathcal{V}$ is n-dimensional vector space over an infinite field $\mathbb{F}$, and let $\{\mathcal{W}_j\}$ be a finite collection of distinct m-dimensional subspaces.

a.   Prove that no $\mathcal{W}_j$ is contained in the union of the others.

b.   Prove that there is a subspace $\mathcal{U} \subset \mathcal{V}$ which is a complement of every $\mathcal{W}_j$.

Hint: See exercise **ex1.2.12**.

∗**ex1.3.17** Assume that any three of the five $\mathbb{R}^3$-vectors $v_j = (x_j, y_j, z_j)$, $j = 1, \ldots, 5$, are linearly independent. Prove that the vectors

$$w_j = (x_j^2, y_j^2, z_j^2, x_j y_j, x_j z_j, y_j z_j)$$

are linearly independent in $\mathbb{R}^6$.

Hint: Find nonzero (a, b, c) such that $ax_j + by_j + cz_j = 0$ for $j = 1, 2$. Find nonzero (d, e, f) such that $dx_j + ey_j + fz_j = 0$ for $j = 3, 4$. Observe (and use) the fact

$$(ax_5 + by_5 + cz_5)(dx_5 + ey_5 + fz_5) \neq 0.$$

1.4   Systems of linear equations

How do we find out if a set $\{v_j\}$, $j = 1, \ldots, m$ of column vectors in $\mathbb{F}_c^n$ is linearly dependent? How do we find out if a vector u belongs to $\mathrm{span}[v_1, \ldots, v_m]$?

Given the vectors $v_j = \begin{bmatrix} a_{1j} \\ \vdots \\ a_{nj} \end{bmatrix}$, $j = 1, \ldots, m$, and $u = \begin{bmatrix} c_1 \\ \vdots \\ c_n \end{bmatrix}$, we ex-

press the conditions $\sum x_j v_j = 0$ for the first question, and $\sum x_j v_j = u$ for the second, in terms of the coordinates.

The first equation leads to the *system of homogeneous linear equations:*

(1.4.1)
$$\begin{aligned}
a_{11}x_1 + \quad \cdots \quad + a_{1m}x_m &= 0 \\
a_{21}x_1 + \quad \cdots \quad + a_{2m}x_m &= 0 \\
\vdots \qquad\qquad \vdots \\
a_{n1}x_1 + \quad \cdots \quad + a_{nm}x_m &= 0
\end{aligned}$$

or,

(1.4.2)
$$\sum_{j=1}^{m} a_{ij}x_j = 0, \quad i = 1,\ldots,n.$$

The second equation gives the *nonhomogeneous system:*

(1.4.3)
$$\sum_{j=1}^{m} a_{ij}x_j = c_i, \quad i = 1,\ldots,n.$$

DEFINITION: The *solution-set* of a system of linear equations, such as (1.4.2) or (1.4.3), is the set of all *m*-tuples $(x_1,\ldots,x_m) \in \mathbb{F}^m$ for which all n equations hold.

To answer the question of the dependence of the v_j's, we need to determine if the solution-set of the system (1.4.2) is *trivial* or not, i.e., if there are solutions other than $(0,\ldots,0)$. To see if $u \in \text{span}[v_1,\ldots,v_m]$, we need to know if the solution-set of (1.4.3) is empty or not.

In both cases we would like to identify the solution-set as completely and as explicitly as possible.

1.4.1 Conversely, beginning with a homogeneous system like (1.4.2) we can rewrite it as

(1.4.4)
$$x_1 \begin{bmatrix} a_{11} \\ \vdots \\ a_{n1} \end{bmatrix} + \cdots + x_m \begin{bmatrix} a_{1m} \\ \vdots \\ a_{nm} \end{bmatrix} = 0$$

and use general properties of vector spaces to draw general conclusions. Our first result depends only on dimension.

Theorem. *A system of n homogeneous linear equations in $m > n$ unknowns has a nontrivial solution.*

PROOF: The m columns in (1.4.4) are elements of the n-dimensional space $\mathbb{F}_c^n$. If $m > n$, then they are dependent, so (1.4.4) has a nontrivial solution.   ◀

Remark: As noted in example *a* of subsection 1.2.3, the set of solutions of a homogeneous system of equations form a subspace of $\mathbb{F}^m$. The theorem above shows that if the number of variables is greater than the number of equations, then the dimension of the subspace of solutions is at least 1. See **ex1.4.1** for a refinement.

Similarly, a nonhomogeneous system like (1.4.3) can be rewritten in the form

(1.4.5)
$$x_1 \begin{bmatrix} a_{11} \\ \vdots \\ a_{n1} \end{bmatrix} + \cdots + x_m \begin{bmatrix} a_{1m} \\ \vdots \\ a_{nm} \end{bmatrix} = \begin{bmatrix} c_1 \\ \vdots \\ c_n \end{bmatrix}.$$

It is then clear that the system given by (1.4.3) has a solution if and only if

the column $\begin{bmatrix} c_1 \\ \vdots \\ c_n \end{bmatrix}$ is in the span of the columns $\begin{bmatrix} a_{1j} \\ \vdots \\ a_{nj} \end{bmatrix}$, $j = 1, \ldots, m$.

1.4.2   The classical approach to solving systems of linear equations is by *Gaussian elimination*—an algorithm for replacing the given system by an *equivalent* system that can be solved easily. We need some terminology:

DEFINITION: The systems

(1.4.6)

$(\mathfrak{A})$ $\displaystyle\sum_{j=1}^{m} a_{ij} x_j = c_i, \quad i = 1, \ldots, k,$

$(\mathfrak{B})$ $\displaystyle\sum_{j=1}^{m} b_{ij} x_j = d_i, \quad i = 1, \ldots, l,$

are *equivalent* if they have the same solution-set (in $\mathbb{F}^m$).

The matrices

$$
A = \begin{bmatrix} a_{11} & \cdots & a_{1m} \\ a_{21} & \cdots & a_{2m} \\ \vdots & \cdots & \vdots \\ a_{k1} & \cdots & a_{km} \end{bmatrix}
\quad \text{and} \quad
A_{aug} = \begin{bmatrix} a_{11} & \cdots & a_{1m} & c_1 \\ a_{21} & \cdots & a_{2m} & c_2 \\ \vdots & \cdots & \vdots & \vdots \\ a_{k1} & \cdots & a_{km} & c_k \end{bmatrix}
$$

are called the *coefficient matrix*, or simply the *matrix*, and the *augmented matrix* of the system ($\mathfrak{A}$). The augmented matrix is obtained from the matrix by appending the column of *values* (i.e., the right-hand sides of the equations in the system) as a (new) last column.

The augmented matrix contains all the information of the system ($\mathfrak{A}$). Any $k \times (m+1)$ matrix is the augmented matrix of a system of linear equations in m unknowns.

1.4.3 Row equivalence of matrices.

DEFINITION: The *row space* of a matrix $A \in \mathcal{M}(k,m)$ is the subspace of $\mathbb{F}_r^m$ spanned by the rows of A. The dimension of this space is called the *row rank* of A.

The matrices

$$
(1.4.7) \quad \begin{bmatrix} a_{11} & \cdots & a_{1m} \\ a_{21} & \cdots & a_{2m} \\ \vdots & \cdots & \vdots \\ a_{k1} & \cdots & a_{km} \end{bmatrix}
\quad \text{and} \quad
\begin{bmatrix} b_{11} & \cdots & b_{1m} \\ b_{21} & \cdots & b_{2m} \\ \vdots & \cdots & \vdots \\ b_{l1} & \cdots & b_{lm} \end{bmatrix}
$$

are *row equivalent* if their rows span the same subspace of $\mathbb{F}_r^m$; equivalently: if each row of either matrix is a linear combination of the rows of the other. Row equivalent matrices clearly have the same row rank.

Proposition. *Two systems of linear equations in m unknowns*

$$
(\mathfrak{A}) \quad \sum_{j=1}^{m} a_{ij} x_j = c_i, \quad i = 1, \dots, k,
$$

$$
(\mathfrak{B}) \quad \sum_{j=1}^{m} b_{ij} x_j = d_i, \quad i = 1, \dots, l,
$$

are equivalent if their respective augmented matrices are row equivalent.

PROOF: Assume that the augmented matrices are row equivalent.
If $(x_1, \ldots, x_m)$ is a solution for system $(\mathfrak{A})$ and

$$(b_{i1}, \ldots, b_{im}, d_i) = \sum \alpha_{i,k}(a_{k1}, \ldots, a_{km}, c_k),$$

then

$$\sum_{j=1}^{m} b_{ij}x_j = \sum_{k,j} \alpha_{i,k}a_{kj}x_j = \sum_{k} \alpha_{i,k}c_k = d_i$$

and $(x_1, \ldots, x_m)$ is a solution for system $(\mathfrak{B})$. ◀

1.4.4   Reduced-row-echelon form. We come now to *Gauss-Jordan elimination*. The equivalent system that is easier to solve is obtained by reducing the augmented matrix of the system to one in *reduced-row-echelon form*.

DEFINITION: A matrix $A \in \mathcal{M}(k,m)$ is in *reduced-row-echelon form* if the following conditions are satisfied:

rref–1   The first q rows of A are linearly independent in $\mathbb{F}^m$, and the remaining $k - q$ rows are zero.

rref–2   There are integers $1 \le l_1 < l_2 < \cdots < l_q \le m$ such that for $j \le q$, the first nonzero entry in the j'th row is 1, occuring in the l_j'th column.

rref–3   The entry 1 in row j is the only nonzero entry in the l_j column.

One can rephrase the last three conditions as: The l_j'th columns, called *pivot* columns, are the first q elements of the standard basis of $\mathbb{F}_c^k$; every other column is a linear combination of the pivot columns that precede it.

Theorem. *Every matrix is row-equivalent to a matrix in reduced-row-echelon form. Furthermore, the reduced-row-echelon form of a matrix is unique.*

PROOF: We describe an algorithm that uses *elementary row operations* to reduce an arbitrary matrix A to a row-equivalent matrix in reduced-row-echelon form.

The *elementary row operations* are:

a. Reordering (i.e., permuting) the rows;

b. Multiplying a row by a nonzero constant;

c. Adding a multiple of one row to another.

These operations do not change the span of the rows so that the row equivalence class of the matrix is maintained. (We shall return later, in exercise **ex2.3.13**, to express these operations as matrix multiplications.)

If $A = 0$, there is nothing to prove, and we assume that $A \neq 0$. Denote the row rank of A by q. Let l_1 be the index of the first column that is not zero.

Reorder the rows so that $a_{1,l_1} \neq 0$, and multiply the first row by a_{1,l_1}^{-1}.

For every $j > 1$, subtract the first row multiplied by a_{j,l_1} from the j'th row.

Now all the columns before l_1 are zero and column l_1 has 1 in the first row, and zero elsewhere.

If the row rank q is 1, all the entries below the first row are now zero and we are done. Otherwise let l_2 be the index of the first column that has a nonzero entry in a row beyond the first. Notice that $l_2 > l_1$. Keep the first row in its place, reorder the remaining rows so that $a_{2,l_2} \neq 0$, and multiply the second row[6] by a_{2,l_2}^{-1}.

For every $j \neq 2$, subtract the second row multiplied by a_{j,l_2} from the j'th row.

Repeat the sequence of steps a total of q times. The first q rows, $\mathbf{r}_1, \ldots, \mathbf{r}_q$, are (now) independent: a combination $\sum c_j \mathbf{r}_j$ has entry c_j in the l_j'th place, and can be zero only if $c_j = 0$ for all j.

If there is a nonzero entry beyond the current q'th row, necessarily beyond the l_q'th column, we could continue and get a row independent of the first q, contradicting the definition of q. Thus, after q steps, all the rows beyond the q'th are zero.

---

[6]We keep referring to the entries of the successively modified matrix as a_{ij}.

Uniqueness of the reduced-row-echelon-form of a matrix is left as exercise **ex1.4.4**.   ◄

Observe that the scalars used in the process belong to the smallest field that contains all the coefficients of A.

1.4.5   If A and A_{aug} are the matrix and the augmented matrix of a system ($\mathfrak{A}$) and we apply the algorithm of the previous subsection to both, we observe that since the augmented matrix has the additional column on the right-hand side, the first q (the row rank of A) steps in the algorithm for either A or A_{aug} are identical. Having done q repetitions, A is in reduced-row-echelon form, while A_{aug} may or may not be. If the row rank of A_{aug} is q, then the algorithm for A_{aug} ends as well; otherwise we have $l_{q+1} = m + 1$, and the reduced-row-echelon form for the augmented matrix is the same as that of A but with an added row and an added pivot column, both having 0 for all but the last entries, and 1 for the last entry. In the latter case, the system corresponding to the row-reduced augmented matrix has as its last equation $0 = 1$ and the system has no solutions.

On the other hand, if the row rank of the augmented matrix is the same as that of A, the reduced-row-echelon form of the augmented matrix is an augmentation of the reduced-row-echelon form of A. In this case we assign arbitrary values to the so-called *free variables*, i.e., the variables x_i, $i \neq l_j$, $j = 1, \dots, q$. We then move the corresponding terms to the right-hand side and, writing C_j for their sum, we obtain

$$(1.4.8) \qquad\qquad x_{l_j} = C_j, \quad j = 1, \dots, q.$$

Theorem. *A necessary and sufficient condition for the system ($\mathfrak{A}$) to have solutions is that the row ranks of the matrix and of the augmented matrix of the system be equal.*

The discussion preceding the statement of the theorem not only proves the theorem but offers a concrete way to solve the system. The unknowns are now split into two groups, q pivot variables and $m - q$ free ones. We have "$m - q$ degrees of freedom": the $m - q$ free unknowns become free parameters that can be assigned arbitrary values, and these values determine the pivot unknowns uniquely.

Remark: Notice that the split into pivot and free unknowns depends on the specific definition of reduced-row-echelon form; counting the columns in a different order may result in a different split, though the number q of pivot variables would be the same, equal to the row rank of A. For example, for the "system" of one equation with two unknowns $x + y = 1$, either x or y can be chosen freely, thereby determining the other.

Corollary. *A linear system of n equations in n unknowns with matrix A has solutions for all augmented matrices if and only if the only solution of the corresponding homogeneous system is the trivial solution.*

PROOF: The condition on the homogeneous system amounts to "the rows of A are independent", and no added columns can increase the row rank. ◄

1.4.6 DEFINITION: The *column space* of a matrix $A \in \mathcal{M}(k,m)$ is the subspace of $\mathbb{F}_c^k$ spanned by the columns of A. The dimension of this space is called the *column rank* of A.

Theorem. *The column rank of a matrix A is equal to its row rank.*

PROOF: Linear relations between columns of A are solutions of the homogeneous system given by A. If B is row-equivalent to A, the columns of A and B have the same set of linear relations (see Proposition 1.4.3). In particular, if the row rank of A is q and B is in reduced-row-echelon form, then the q pivot columns in B are independent, and every other column is a linear combination of these. ◄

We refer to the common value of the row and column ranks of A simply as *the rank of A* and denote it by $\rho(A)$.

1.4.7 DEFINITION. A *submatrix* of a matrix A is a matrix B obtained by deleting from A some rows and some columns.[7]

B is a *principal submatrix* of a square matrix A if the set of indices of the deleted columns is the same as that of the deleted rows.

---

[7]We use the word *some* to mean *a nonnegative number of*, that is, *some or none*.

A simple corollary of Theorem 1.4.6 is the following proposition.

Proposition. *If A is a matrix and $\rho(A) = k$, then there is a $k \times k$ submatrix B of A such that $\rho(B) = k$.*

See **ex1.4.12** for a refinement.

EXERCISES FOR SECTION 1.4

ex1.4.1 Show that if $m > n$, then the dimension of the space of solutions of a homogeneous system of n linear equations in m variables is at least $m - n$.

ex1.4.2 Prove Proposition 1.4.7, and show an example of a 3×3 matrix of rank 2 without a principal submatrix of rank 2. (So that the second part of the theorem is false without some assumption of symmetry.)

ex1.4.3 Identify the matrix $A \in \mathcal{M}(n)$ of row rank n that is in reduced-row-echelon form.

ex1.4.4 Show that if $A, B \in \mathcal{M}(k,m)$ are both in reduced-row-echelon form, then either $A = B$, or A and B are not row equivalent. Conclude that the reduced-row-echelon form of a matrix is unique.

Hint: Consider the solution sets of the two homogeneous systems with coefficient matrices A and B, respectively. Show that these must be different if $A \neq B$.

ex1.4.5 A system of linear equations with rational coefficients that has a solution in $\mathbb{C}$, has a solution in $\mathbb{Q}$. Equivalently, vectors in $\mathbb{Q}^n$ that are linearly dependent over $\mathbb{C}$ are rationally dependent.

Hint: The last sentence of subsection 1.4.4.

ex1.4.6 A system of linear equations with rational coefficients, has the same number of degrees of freedom over $\mathbb{Q}$ as it does over $\mathbb{C}$.[8]

ex1.4.7 A subset $\mathcal{A}$ of a vector space $\mathcal{V}$ is called an *affine subspace* if $\mathcal{A}$ is the translate of a subspace $\mathcal{W}$ of $\mathcal{V}$, i.e., $\mathcal{A} = \{v_0 + w : w \in \mathcal{W}\}$. We call the subspace $\mathcal{W}$ the *corresponding subspace* to $\mathcal{A}$. (Thus a *line* in $\mathcal{V}$ is a translate of a one-dimensional subspace.)

Prove that a set $\mathcal{A} \subset \mathcal{V}$ is an affine subspace if and only if $\sum a_j u_j \in \mathcal{A}$ for all choices of $u_1, \ldots, u_k \in \mathcal{A}$, and scalars $a_j, j = 1, \ldots, k$ such that $\sum a_j = 1$.

---

[8]See 1.4.5.

ex1.4.8 Let $\mathscr{A} \subset \mathscr{V}$ be an affine subspace and $u_0 \in \mathscr{A}$. Prove that the set $\mathscr{A} - u_0 = \{u - u_0 : u \in \mathscr{A}\}$ is the corresponding subspace of $\mathscr{A}$ in $\mathscr{V}$. Show that the corresponding subspace $\mathscr{A} - u_0$ does not depend on the choice of u_0 in $\mathscr{A}$.

ex1.4.9 Show that the solution-set of a system of k linear equations in m unknowns is an affine subspace of $\mathbb{F}^m$. What is its corresponding subspace?

ex1.4.10 A column $v_j = \begin{bmatrix} a_{1j} \\ \vdots \\ a_{kj} \end{bmatrix}$ of a matrix $A = \begin{bmatrix} a_{11} & \cdots & a_{1m} \\ \vdots & \cdots & \vdots \\ a_{k1} & \cdots & a_{km} \end{bmatrix}$ is called a pivot column if j is the index of a pivot column in the reduced-row-echelon form of A. Show that v_j is a pivot column of A if and only if it is linearly independent of the columns v_i, $i < j$.

ex1.4.11 Denote by $B = \begin{bmatrix} b_{11} & \cdots & b_{1m} \\ \vdots & \cdots & \vdots \\ b_{k1} & \cdots & b_{km} \end{bmatrix}$ the reduced-row-echelon form of the matrix A in the previous problem. Let $l_1 < l_2, \ldots$ be the indices of the pivot columns in B and i the index of another column. Prove that

$$(1.4.9) \qquad\qquad v_i = \sum_{l_j < i} b_{ji} v_{l_j},$$

where, as in the previous problem, $v_1, \ldots, v_m$ are the columns of A.

ex1.4.12 Show that if the matrix A in Proposition 1.4.7 is symmetric or skew-symmetric, then the submatrix B may be taken to be a *principal* submatrix.

ex1.4.13 What is the reduced-row-echelon form of the 7×6 matrix A, if its columns C_j, $j = 1, \ldots, 6$, satisfy the following conditions:

a. $C_1 \neq 0$;

b. $C_2 = 3C_1$;

c. C_3 is not a (scalar) multiple of C_1;

d. $C_4 = C_1 + 2C_2 + 3C_3$;

e. $C_5 = 6C_3$;

f. C_6 is not in the span of C_2 and C_3.

*ex1.4.14 Given polynomials $P_1 = \sum_0^n a_j x^j$, $P_2 = \sum_0^m b_j x^j$, and $S = \sum_0^l s_j x^j$ of degrees n, m, and $l < n + m$ respectively, suppose that we want to find polynomials $q_1 = \sum_0^{m-1} c_j x^j$ and $q_2 = \sum_0^{n-1} d_j x^j$ such that

$$(1.4.10) \qquad\qquad P_1 q_1 + P_2 q_2 = S.$$

The polynomial equation (1.4.10) is equivalent to the system of $m + n$ linear equations,

$$(1.4.11) \qquad \sum_{j+k=l} a_j c_k + \sum_{r+t=l} b_r d_t = s_l, \quad l = 0, \ldots, n+m-1,$$

where the unknowns are the coefficients $c_{m-1}, \ldots, c_0$ of q_1, and $d_{n-1}, \ldots, d_0$ of q_2. The coefficient matrix for this system is known as the *Sylvester matrix* of (P_1, P_2).

Write the matrix of the system for $n = 3$ and $m = 2$.

*ex1.4.15 The associated homogeneous system to the system (1.4.11) corresponds to the case $S = 0$. Show that it has a nontrivial solution if and only if P_1 and P_2 have a nontrivial common factor. (You may assume the unique factorization theorem; see A.6.3 in the appendix.) What is the rank of the Sylvester matrix if the degree of $\gcd(P_1, P_2)$ is r?

*1.5   Normed finite-dimensional linear spaces

1.5.1   A norm on a real or complex vector space $\mathcal{V}$ is a nonnegative function $v \mapsto \|v\|$ that satisfies the conditions

a.   Positivity: $\|0\| = 0$ and if $v \neq 0$ then $\|v\| > 0$.

b.   Homogeneity: $\|av\| = |a|\|v\|$ for scalars a and vectors v.

c.   The triangle inequality: $\|v + u\| \leq \|v\| + \|u\|$.

These properties guarantee that $\delta(v, u) = \|v - u\|$ is a metric on the space, the metric *defined* or *induced* by the norm; and with a metric one can use tools and notions from point-set topology such as limits, continuity, convergence, infinite series, etc.

A *normed vector space* is a vector space endowed with a norm. Since $\mathbb{R}$ and $\mathbb{C}$ are *complete* metric spaces, so is every normed finite-dimensional real or complex vector space.

1.5.2 If $\mathscr{V}$ and $\mathscr{W}$ are isomorphic real or complex finite-dimensional spaces and S is an isomorphism of $\mathscr{V}$ onto $\mathscr{W}$, then a norm $\|\cdot\|_{\mathscr{W}}$ on $\mathscr{W}$ can be "carried back" to $\mathscr{V}$ by defining $\|v\|_{\mathscr{V}} = \|Sv\|_{\mathscr{W}}$. This implies that all possible norms on a real n-dimensional space are copies of norms on $\mathbb{R}^n$, and all norms on a complex n-dimensional space are copies of norms on $\mathbb{C}^n$.

A finite-dimensional $\mathscr{V}$ can be endowed with many different norms; yet, all these norms are *equivalent* in the following sense (see **ex1.5.1**):

DEFINITION: The norms $\|\cdot\|_1$ and $\|\cdot\|_2$ are *equivalent* if there is a positive constant C such that for all $v \in \mathscr{V}$,

$$C^{-1}\|v\|_1 \le \|v\|_2 \le C\|v\|_1.$$

Metrics δ_1, δ_2, defined by equivalent norms $\|\cdot\|_1$ and $\|\cdot\|_2$, are equivalent: for $v, u \in \mathscr{V}$,

$$C^{-1}\delta_1(v,u) \le \delta_2(v,u) \le C\delta_1(v,u),$$

which means that they define the same topology—the familiar topology of $\mathbb{R}^n$ or $\mathbb{C}^n$.

EXERCISES FOR SECTION 1.5

ex1.5.1 Let $\mathscr{V}$ be an n-dimensional real or complex vector space, and let $\mathbf{v} = \{v_1,\dots,v_n\}$ be a basis for $\mathscr{V}$. Write

$$\left\|\sum a_j v_j\right\|_{\mathbf{v},1} = \sum |a_j| \qquad \text{and} \qquad \left\|\sum a_j v_j\right\|_{\mathbf{v},\infty} = \max |a_j|.$$

Prove:

a. $\|\cdot\|_{\mathbf{v},1}$ and $\|\cdot\|_{\mathbf{v},\infty}$ are norms on $\mathscr{V}$, and

(1.5.1) $$\|\cdot\|_{\mathbf{v},\infty} \le \|\cdot\|_{\mathbf{v},1} \le n\|\cdot\|_{\mathbf{v},\infty}.$$

b. If $\|\cdot\|$ is *any* norm on $\mathscr{V}$ then, for all $v \in \mathscr{V}$,

(1.5.2) $$\|v\|_{\mathbf{v},1} \max\|v_j\| \ge \|v\|.$$

c. Let $\|\cdot\|_j$, $j = 1,2$, be norms on $\mathscr{V}$, and δ_j the induced metrics. Let $\{v_n\}_{n=0}^{\infty}$ be a sequence in $\mathscr{V}$. Prove that if $\delta_1(v_n, v_0) \to 0$, then $\delta_2(v_n, v_0) \to 0$.

d. Show that if $\|\cdot\|$ is an arbitrary norm on $\mathscr{V}$, then the function $f : \mathbb{F}^n \to \mathbb{R}$ (where $\mathbb{F}$ is the field of scalars of $\mathscr{V}$, either $\mathbb{R}$ or $\mathbb{C}$), defined by

$$f(a_1, \ldots, a_n) = \left\| \sum a_j v_j \right\|,$$

is continuous on $\mathbb{F}^n$. Conclude that f attains a strictly positive minimum on the set $B = \{(a_1, \ldots, a_n) : \sum |a_j| = 1\} \subset \mathbb{F}^n$.

Hint: Show that the triangle inequality implies that $\big| \|v\| - \|u\| \big| \leq \|v - u\|$, then use part **b** and the compactness of B.

e. Conclude that all norms on $\mathscr{V}$ are equivalent to $\|\cdot\|_{\mathbf{v},1}$.

ex1.5.2 Let $\{v_n\}_{n=0}^{\infty}$ be bounded in $\mathscr{V}$. Prove that $\sum_0^{\infty} z^n v_n$ converges for every z such that $|z| < 1$.

Hint: Prove that the partial sums form a Cauchy sequence in the metric defined by the norm.

ex1.5.3 Let $\mathscr{V}$ be n-dimensional real or complex normed vector space. The *unit ball* in $\mathscr{V}$ is the set

$$B_1 = \{v \in \mathscr{V} : \|v\| \leq 1\}.$$

Prove that B_1 is

a. *Convex:* If $v, u \in B_1$, $0 \leq a \leq 1$, then $av + (1 - a)u \in B_1$.

b. *Bounded:* For every $v \in \mathscr{V}$, there exists a (positive) constant λ such that $cv \notin B$ for $|c| > \lambda$.

c. *Circularly symmetric, centered at* 0: If $v \in B$ and $|a| = 1$ then $av \in B$.

   Notice that convexity and circular symmetry together imply that if $v \in B$ and $|a| \leq 1$ then $av \in B$.

ex1.5.4 Let $\mathscr{V}$ be n-dimensional real or complex vector space, and let B be a bounded circularly symmetric convex set centered at 0. Define

$$\|u\| = \inf\{a > 0 : a^{-1}u \in B\}.$$

Prove that this defines a norm on $\mathscr{V}$, and the unit ball for this norm is the given set B.

ex1.5.5 Describe a norm $\|\ \|_0$ on $\mathbb{R}^3$ such that the standard unit vectors have norm 1 while $\|(1, 1, 1)\|_0 < \frac{1}{100}$.

Chapter 2

Linear Operators and Matrices

2.1   Linear operators

2.1.1   Let $\mathcal{V}$ and $\mathcal{W}$ be vector spaces over the same field $\mathbb{F}$. Recall definition 1.2.2:

DEFINITION: A map $T : \mathcal{V} \to \mathcal{W}$ is *linear* if for all vectors $v_j \in \mathcal{V}$ and scalars a_j,

$$(2.1.1) \qquad T(a_1 v_1 + a_2 v_2) = a_1 T v_1 + a_2 T v_2.$$

Induction on the the number k of summands proves that equation (2.1.1) implies

$$(2.1.2) \qquad T\Big(\sum_{j=1}^{k} a_j v_j\Big) = \sum_{j=1}^{k} a_j T v_j$$

for any number of summands.

Linear maps are also called *linear operators, linear transformations, homomorphisms,* etc. The adjective "linear" is sometimes assumed implicitly. The term that we use most is *operator*, and we *always* mean by that a *linear operator*.

EXAMPLES:

a.   The identity map I of $\mathcal{V}$ onto itself, defined by $Iv = v$ for $v \in \mathcal{V}$. For $\lambda \in \mathbb{F}$ the multiplication operator $v \mapsto \lambda v$ is a linear operator on $\mathcal{V}$. We denote it by λI or simply by λ.

b.   Let $\mathcal{V}$ be the space of all continuous, 2π-periodic functions on the line. For every x_0 define T_{x_0}, the *translation by x_0*:

$$T_{x_0} : f(x) \mapsto f_{x_0}(x) = f(x - x_0).$$

c. The *transpose*,

(2.1.3) $A = \begin{bmatrix} a_{11} & \cdots & a_{1m} \\ a_{21} & \cdots & a_{2m} \\ \vdots & \cdots & \vdots \\ a_{n1} & \cdots & a_{nm} \end{bmatrix} \mapsto A^{Tr} = \begin{bmatrix} a_{11} & \cdots & a_{n1} \\ a_{12} & \cdots & a_{n2} \\ \vdots & \cdots & \vdots \\ a_{1m} & \cdots & a_{nm} \end{bmatrix},$

which maps $\mathcal{M}(n,m;\mathbb{F})$ onto $\mathcal{M}(m,n;\mathbb{F})$.

d. Differentiation on $C^{\infty}([0,1])$ (the complex vector space of infinitely differentiable complex-valued functions on $[0,1]$):

(2.1.4) $D\colon f \mapsto f' = \dfrac{df}{dx}.$

e. Integration on $C([0,1])$ (the complex vector space of continuous complex-valued functions on $[0,1]$),

(2.1.5) $\mathsf{Int}\colon f \mapsto \displaystyle\int_0^1 f(x)\,dx,$

is a linear map from $C([0,1])$ into $\mathbb{C}$.

f. If $\mathcal{V} = \mathcal{W} \oplus \mathcal{U}$, then every $v \in \mathcal{V}$ has a unique representation $v = w + u$ with $w \in \mathcal{W}$, $u \in \mathcal{U}$; w is *the component of v in $\mathcal{W}$*, and u the component in $\mathcal{U}$. The map $\pi_1\colon v \mapsto w$, mapping each vector on its component in $\mathcal{W}$, is the identity on $\mathcal{W}$ and maps $\mathcal{U}$ to $\{0\}$. It is called the *projection of $\mathcal{V}$ on $\mathcal{W}$ along $\mathcal{U}$*.

The operator π_1 is linear since, if $v = w + u$ and $v_1 = w_1 + u_1$, then $av + bv_1 = (aw + bw_1) + (au + bu_1)$. Now $aw + bw_1 \in \mathcal{W}$, and $au + bu_1 \in \mathcal{U}$, so that $\pi_1(av + bv_1) = a\pi_1 v + b\pi_1 v_1$.

Similarly, $\pi_2\colon v \mapsto u$ is called the *projection of $\mathcal{V}$ on $\mathcal{U}$ along $\mathcal{W}$*. The maps π_1 and π_2 are referred to as the *projections corresponding to the direct sum decomposition*.

g. *Restricting* a linear operator. Let $\mathcal{V}$ and $\mathcal{W}$ be vector spaces over $\mathbb{F}$ and let $T\colon \mathcal{V} \to \mathcal{W}$ be linear. If $\mathcal{V}_1 \subset \mathcal{V}$ is a subspace, we define the *restriction* $T_{\mathcal{V}_1}$ of T to $\mathcal{V}_1$ by setting, for $v \in \mathcal{V}_1$, $T_{\mathcal{V}_1} v = Tv$. Then $T_{\mathcal{V}_1}$ is a linear operator from $\mathcal{V}_1$ into $\mathcal{W}$.

2.1.2 Let $\mathscr{V}$ and $\mathscr{W}$ be vector spaces over the same field $\mathbb{F}$. Assume that $\mathscr{V}$ is finite-dimensional, and let $\{v_1, \ldots, v_n\}$ be a basis. For any choice of vectors $w_1, \ldots, w_n$ in $\mathscr{W}$, the map $v_j \mapsto w_j$, $j = 1, \ldots, n$ extends uniquely to a linear operator T from $\mathscr{V}$ to $\mathscr{W}$, defined by:

$$(2.1.6) \qquad T: \sum a_j v_j \mapsto \sum a_j w_j.$$

In other words:

Theorem. *A linear operator from a finite-dimensional space $\mathscr{V}$ to a space $\mathscr{W}$ is completely determined by the values it assigns to the elements of a given basis of $\mathscr{V}$, and these values can be arbitrary vectors in $\mathscr{W}$.*

PROOF: The map T, as defined by (2.1.6), is clearly linear, and, by (2.1.2), a *linear* operator that agrees with T on $\{v_1, \ldots, v_n\}$ must agree with it on the entire space. ◀

Consider, for example, linear operators from $\mathbb{F}^n_{\mathbf{c}}$ into $\mathbb{F}^m_{\mathbf{c}}$. We choose a basis $\mathbf{v} = \{v_1, \ldots, v_n\}$ of $\mathbb{F}^n_{\mathbf{c}}$ and observe that, given this choice and given a matrix $A = (a_{j,i}) \in \mathscr{M}(m, n, \mathbb{F})$, we can define an operator T_A by setting $T_A v_i$ to be the i'th column of the matrix:

$$(2.1.7) \qquad T_A v_i = \begin{bmatrix} a_{1,i} \\ \vdots \\ a_{m,i} \end{bmatrix}, \quad \text{so that} \quad T_A\left(\sum c_i v_i\right) = \begin{bmatrix} \sum c_i a_{1,i} \\ \vdots \\ \sum c_i a_{m,i} \end{bmatrix};$$

and hence T_A is completely determined by the explicit choice of the basis $\mathbf{v}$, the matrix A, and the (implicit) choice of the standard basis in $\mathbb{F}^m_{\mathbf{c}}$.

2.1.3 Given vector spaces $\mathscr{V}$ and $\mathscr{W}$, we denote the space of all linear operators from $\mathscr{V}$ into $\mathscr{W}$ by $\mathscr{L}(\mathscr{V}, \mathscr{W})$. Another common notation is $HOM(\mathscr{V}, \mathscr{W})$. The two most important cases in what follows are: $\mathscr{W} = \mathscr{V}$, and $\mathscr{W} = \mathbb{F}$, the field of scalars.

• When $\mathscr{W} = \mathscr{V}$, we write $\mathscr{L}(\mathscr{V})$ instead of $\mathscr{L}(\mathscr{V}, \mathscr{V})$.

• When $\mathscr{W}$ is the underlying field, we refer to the linear maps as *linear functionals* or *linear forms* on $\mathscr{V}$. Instead of $\mathscr{L}(\mathscr{V}, \mathbb{F})$ we write $\mathscr{V}^*$, and refer to it as *the dual space* of $\mathscr{V}$ (see Chapter 3).

2.1.4   Coordinates. If $\mathscr{V}$ is a finite-dimensional space, every basis $\mathbf{v} = \{v_1, \ldots, v_n\}$ of $\mathscr{V}$ defines an isomorphism $\mathsf{C}_{\mathbf{v}}$ of $\mathscr{V}$ onto $\mathbb{F}^n$ by

$$(2.1.8) \qquad \mathsf{C}_{\mathbf{v}}: v = \sum a_j v_j \mapsto \begin{bmatrix} a_1 \\ \vdots \\ a_n \end{bmatrix} = \sum a_j e_j.$$

The vector $\mathsf{C}_{\mathbf{v}} v$ is the *coordinates vector* of v relative to the basis $\mathbf{v}$. Notice that this is a special case of (2.1.6) above: we map the basis elements v_j on the corresponding elements e_j of the standard basis, and extend by linearity.

2.1.5 $\mathscr{L}(\mathscr{V}, \mathscr{W})$ as a vector space. We define the sum of linear maps $T, S \in \mathscr{L}(\mathscr{V}, \mathscr{W})$ and the multiple of a linear map by a scalar, as follows: for every $v \in \mathscr{V}$,

$$(2.1.9) \qquad (T + S)v = Tv + Sv, \quad (aT)v = a(Tv).$$

Observe that $(T + S)$ and aT, as defined, are linear maps from $\mathscr{V}$ to $\mathscr{W}$, that is, are elements of $\mathscr{L}(\mathscr{V}, \mathscr{W})$. A straightforward verification, left to the reader, shows that the *addition* and *multiplication by scalars* as defined by (2.1.9) satisfy the requirements imposed on addition and multiplication by scalars in the definition of a vector space. The addition is associative, commutative, and distributive with respect to multiplication by scalars. It follows that if $\mathscr{V}$ and $\mathscr{W}$ are vector spaces over $\mathbb{F}$, then, with addition and the multiplication by scalars as defined by (2.1.9), $\mathscr{L}(\mathscr{V}, \mathscr{W})$ is a vector space over $\mathbb{F}$.

Proposition. *If both $\mathscr{V}$ and $\mathscr{W}$ are finite-dimensional, then so is $\mathscr{L}(\mathscr{V}, \mathscr{W})$, and* $\dim \mathscr{L}(\mathscr{V}, \mathscr{W}) = \dim \mathscr{V} \dim \mathscr{W}$.

The proof is left to the reader as exercise **ex2.1.4** below.

EXERCISES FOR SECTION 2.1

ex2.1.1 Prove (verify) that the operator T defined in (2.1.7) is linear.

ex2.1.2 Show that if a set $A \subset \mathscr{V}$ is linearly dependent and $T \in \mathscr{L}(\mathscr{V}, \mathscr{W})$, then TA is linearly dependent in $\mathscr{W}$.

ex2.1.3 Prove that an injective map $T \in \mathcal{L}(\mathcal{V}, \mathcal{W})$ is an isomorphism if and only if it maps *some* basis of $\mathcal{V}$ onto a basis of $\mathcal{W}$, and this is the case if and only if it maps *every* basis of $\mathcal{V}$ onto a basis of $\mathcal{W}$.

ex2.1.4 Let $\mathcal{V}$ and $\mathcal{W}$ be finite-dimensional with bases $\{v_1, \ldots, v_n\}$ and $\{w_1, \ldots, w_m\}$ respectively. Let $\varphi_{ij} \in \mathcal{L}(\mathcal{V}, \mathcal{W})$ be defined by $\varphi_{ij} v_i = w_j$ and $\varphi_{ij} v_k = 0$ for $k \neq i$. Prove that $\{\varphi_{ij} : 1 \leq i \leq n, \ 1 \leq j \leq m\}$ is a basis for $\mathcal{L}(\mathcal{V}, \mathcal{W})$.

ex2.1.5 Let $\mathcal{V}$ and $\mathcal{W}$ be finite-dimensional with bases $\{v_1, \ldots, v_n\}$ and $\{w_1, \ldots, w_m\}$ respectively. Prove that $\mathcal{L}(\mathcal{V}, \mathcal{W})$ is isomorphic to $\mathcal{M}(m, n, \mathbb{F})$.

2.2   Operator multiplication

2.2.1   For $T \in \mathcal{L}(\mathcal{V}, \mathcal{W})$ and $S \in \mathcal{L}(\mathcal{W}, \mathcal{U})$ we define the product ST in $\mathcal{L}(\mathcal{V}, \mathcal{U})$ by *composition,* that is: $(ST)v = S(Tv)$.

As defined, ST clearly maps $\mathcal{V}$ into $\mathcal{U}$; it is a *linear operator* since

$$(2.2.1) \quad ST(a_1 v_1 + a_2 v_2) = S(a_1 T v_1 + a_2 T v_2) = a_1 ST v_1 + a_2 ST v_2.$$

2.2.2   Of particular interest is the special case where $\mathcal{V} = \mathcal{W} = \mathcal{U}$ so that T, S, and TS are all in $\mathcal{L}(\mathcal{V})$.

Proposition. *With the product ST defined above, $\mathcal{L}(\mathcal{V})$ is an algebra[1] over $\mathbb{F}$.*

PROOF: The claim is that the product is associative and distributive (with the addition defined by (2.1.9)). That is: if $R, S, T \in \mathcal{L}(\mathcal{V})$ then

$$(2.2.2) \quad \begin{aligned} R(ST) &= (RS)T, \\ R(S+T) &= RS + RT, \quad \text{and} \quad (R+S)T = RT + ST. \end{aligned}$$

This is a straightforward checking, left to the reader. ◀

The algebra $\mathcal{L}(\mathcal{V})$ is *not commutative* unless $\dim \mathcal{V} = 1$, in which case it is simply the underlying field.

---

[1] See the appendix, A.5.6.

The set of automorphisms, i.e., invertible elements in $\mathscr{L}(\mathscr{V})$, is *a group* under multiplication, the *general linear group* of $\mathscr{V}$. It is denoted $\mathbf{GL}(\mathscr{V})$.

DEFINITION: The operators $T, S \in \mathscr{L}(\mathscr{V})$ are *conjugate* if for some $R \in \mathbf{GL}(\mathscr{V})$,

$$(2.2.3) \qquad\qquad T = RSR^{-1}.$$

The relation is symmetric ($S = R^{-1}TR$), reflexive, and transitive—it is an equivalence relation.

2.2.3   Given an operator $T \in \mathscr{L}(\mathscr{V})$, the powers T^j of T are well defined for all integers $j \geq 1$, and we define $T^0 = I$. Since we can take linear combinations of the powers of T, we have $P(T)$ well defined for all polynomials $P \in \mathbb{F}[x]$. Specifically, if $P(x) = \sum a_j x^j$ then $P(T) = \sum a_j T^j$.

We denote

$$(2.2.4) \qquad\qquad \mathscr{P}(T) = \{P(T) : P \in \mathbb{F}[x]\}.$$

$\mathscr{P}(T)$ is a subalgebra of $\mathscr{L}(\mathscr{V})$; it will be the main tool in understanding the way in which T acts on $\mathscr{V}$.

It is convenient to introduce the following terminology.

DEFINITION: A *linear system* is a pair $(\mathscr{V}, T)$ where $\mathscr{V}$ is a vector space and $T \in \mathscr{L}(\mathscr{V})$. When we add adjectives, they apply in the appropriate place, so that a *finite-dimensional system* is a system in which $\mathscr{V}$ is finite-dimensional, while an *invertible system* is one in which T is invertible.

EXERCISES FOR SECTION 2.2

ex2.2.1 Give an example of operators $T, S \in \mathscr{L}(\mathbb{R}^2)$ such that $TS \neq ST$.

ex2.2.2 Prove that, for any $T \in \mathscr{L}(\mathscr{V})$, $\mathscr{P}(T)$ is a commutative subalgebra of $\mathscr{L}(\mathscr{V})$.

ex2.2.3 Let $T \in \mathscr{L}(\mathbb{R}^2)$ be defined by $Te_1 = e_2$ and $Te_2 = e_1$. Show that $\mathscr{P}(T) = \{aT + b : a, b \in \mathbb{R}\}$.[2] What are the invertible elements in $\mathscr{P}(T)$?

---

[2]In this context, b is a shorthand for the operator bI.

ex2.2.4 For $T \in \mathscr{L}(\mathscr{V})$ denote $\mathrm{comm}[T] = \{S : S \in \mathscr{L}(\mathscr{V}), ST = TS\}$, the set of operators that commute with T. Prove that $\mathrm{comm}[T]$ is a subalgebra of $\mathscr{L}(\mathscr{V})$.

ex2.2.5 Verify that $\mathbf{GL}(\mathscr{V})$ is in fact a group.

ex2.2.6 An element $\pi \in \mathscr{L}(\mathscr{V})$ is *idempotent* if $\pi^2 = \pi$. Prove that an idempotent π is a projection onto its range, $\pi\mathscr{V} = \{\pi v : v \in \mathscr{V}\}$, along its kernel, $\ker(\pi) = \{v : \pi v = 0\}$.

2.3 Matrix multiplication

2.3.1 The product $\mathbf{r} \cdot \mathbf{c}$ of a row $\mathbf{r}$ (the $1 \times n$ matrix $\mathbf{r} = [a_1, \ldots, a_n]$) and a column $\mathbf{c}$ (the $n \times 1$ matrix $\mathbf{c} = \begin{bmatrix} b_1 \\ \vdots \\ b_n \end{bmatrix}$) is, by definition, the scalar given by

$$(2.3.1) \qquad \mathbf{r} \cdot \mathbf{c} = \sum a_j b_j.$$

Given $A \in \mathscr{M}(l,m)$ and $B \in \mathscr{M}(m,n)$, we define the *product AB* as the $l \times n$ matrix C whose entries c_{ij} are given by

$$(2.3.2) \qquad c_{ij} = \mathbf{r}_i(A) \cdot \mathbf{c}_j(B) = \sum_k a_{ik} b_{kj}$$

($\mathbf{r}_i(A)$ denotes the i'th row in A, and $\mathbf{c}_j(B)$ denotes the j'th column in B).

Notice that the product is defined only when the number of columns in A (the length of its rows) is the same as the number of rows in B, (the height of its columns).

The product is associative: given $A \in \mathscr{M}(l,m)$, $B \in \mathscr{M}(m,n)$, and $C \in \mathscr{M}(n,p)$, then $AB \in \mathscr{M}(l,n)$ and the product $(AB)C \in \mathscr{M}(l,p)$ is well defined. Similarly, $A(BC)$ is well defined and one checks that $A(BC) = (AB)C$ by verifying that the r, s entry in either is $\sum_{i,j} a_{rj} b_{ji} c_{is}$.

The product is distributive: for $A_j \in \mathscr{M}(l,m)$, $B_j \in \mathscr{M}(m,n)$,

$$(2.3.3) \qquad (A_1 + A_2)(B_1 + B_2) = A_1 B_1 + A_1 B_2 + A_2 B_1 + A_2 B_2,$$

and commutes with multiplication by scalars: $(aA)B = A(aB) = a(AB)$.

Going back to the definition and observing that $A \mapsto \mathbf{r}_i(A)$ and $B \mapsto \mathbf{c}_j(B)$ are linear maps, we have

Proposition. *The map $(A, B) \mapsto AB$, of $\mathcal{M}(l, m) \times \mathcal{M}(m, n)$ to $\mathcal{M}(l, n)$, is linear in B for every fixed A, and linear in A for every fixed B.*

2.3.2   Write the $m \times n$ matrix $A = \left(a_{ij}\right)_{\substack{1 \le i \le m \\ 1 \le j \le n}}$ as a "single row of columns",

$$\begin{bmatrix} a_{11} & \cdots & a_{1n} \\ a_{21} & \cdots & a_{2n} \\ \vdots & \cdots & \vdots \\ a_{m1} & \cdots & a_{mn} \end{bmatrix} = \left[\begin{bmatrix} a_{11} \\ a_{21} \\ \vdots \\ a_{m1} \end{bmatrix} \begin{bmatrix} a_{12} \\ a_{22} \\ \vdots \\ a_{m2} \end{bmatrix} \cdots \begin{bmatrix} a_{1n} \\ a_{2n} \\ \vdots \\ a_{mn} \end{bmatrix} \right] = \left(\mathbf{c}_1(A), \mathbf{c}_2(A), \ldots, \mathbf{c}_n(A) \right),$$

we have, for every n-column $\mathbf{y}$,

$$\begin{bmatrix} a_{11} & \cdots & a_{1n} \\ a_{21} & \cdots & a_{2n} \\ \vdots & \cdots & \vdots \\ a_{m1} & \cdots & a_{mn} \end{bmatrix} \begin{bmatrix} y_1 \\ y_2 \\ \vdots \\ y_n \end{bmatrix} = \left(\mathbf{c}_1(A), \mathbf{c}_2(A), \ldots, \mathbf{c}_n(A) \right) \begin{bmatrix} y_1 \\ y_2 \\ \vdots \\ y_n \end{bmatrix} = \sum_{j=1}^{n} y_j \mathbf{c}_j(A),$$

so that for $\mathbf{y} \in \mathbb{F}_{\mathbf{c}}^n$, every column $A\mathbf{y}$ is a linear combination (with weights y_j) of the columns of A.

Similarly, we can write the $n \times p$ matrix B as a "single column of rows",

$$B = \begin{bmatrix} b_{11} & \cdots & b_{1p} \\ b_{21} & \cdots & b_{2p} \\ \vdots & \cdots & \vdots \\ b_{n1} & \cdots & b_{np} \end{bmatrix} = \begin{bmatrix} [b_{11} & \cdots & b_{1p}] \\ [b_{21} & \cdots & b_{2p}] \\ \vdots \\ [b_{n1} & \cdots & b_{np}] \end{bmatrix} = \begin{bmatrix} \mathbf{r}_1(B) \\ \mathbf{r}_2(B) \\ \vdots \\ \mathbf{r}_n(B) \end{bmatrix},$$

where $\mathbf{r}_i(B)$ is the row $(b_{i,1}, \ldots, b_{i,p}) \in \mathbb{F}_{\mathbf{r}}^p$.

If $[x_1, \ldots, x_n] \in \mathbb{F}_{\mathbf{r}}^n$, then

$$(2.3.4) \quad [x_1, \ldots, x_n] \begin{bmatrix} b_{11} & \cdots & b_{1p} \\ b_{21} & \cdots & b_{2p} \\ \vdots & \cdots & \vdots \\ b_{n1} & \cdots & b_{np} \end{bmatrix} = [x_1, \ldots, x_n] \begin{bmatrix} \mathbf{r}_1(B) \\ \mathbf{r}_2(B) \\ \vdots \\ \mathbf{r}_n(B) \end{bmatrix} = \sum_{i=1}^{n} x_i \mathbf{r}_i(B).$$

Proposition. *Assume $A \in \mathcal{M}(m, n)$, $B \in \mathcal{M}(n, p)$; then every column of the matrix AB is a linear combination of the columns of A, and every row of AB is a linear combination of the rows of B.*

We leave the final verification as an exercise (**ex2.3.2**) to the reader.

2.3.3 If $l = m = n$, matrix multiplication is a product within $\mathcal{M}(n)$.

Proposition. *With the multiplication defined above,* $\mathcal{M}(n)$ *is an algebra over* $\mathbb{F}$. *The matrix* $I = I_n = (\delta_{j,k}) = \sum_1^n E_{ii}$ *is the identity[3] element in* $\mathcal{M}(n)$.

The invertible elements in $\mathcal{M}(n)$ form a group under multiplication, the *general linear group* $\mathbf{GL}(n, \mathbb{F})$.

Theorem. *A matrix* $A \in \mathcal{M}(n)$ *is invertible if and only if* $\rho(A) = n$.

PROOF: Proposition 2.3.2 guarantees that the row rank of BA is no bigger than the row rank of A. If $BA = I$, the row rank of A is at least the row rank of I, which is clearly n.

On the other hand, if A is row equivalent to I, then its reduced-row-echelon form is I, and by exercise **ex2.3.13** below, reduction to reduced-row-echelon form amounts to multiplication on the left by a matrix B, so that $BA = I$, i.e., A has a *left inverse*. This implies that A is invertible (see Exercise **ex2.3.15**). ◀

DEFINITION: The matrices $A, B \in \mathcal{M}(n)$ are *conjugate* (by $\mathbf{GL}(n, \mathbb{F})$) if there exists $C \in \mathbf{GL}(n, \mathbb{F})$ such that

$$(2.3.5) \qquad\qquad A = CBC^{-1}.$$

As for operators, this is an equivalence relation.

EXERCISES FOR SECTION 2.3

ex2.3.1 Let **r** be the $1 \times n$ matrix all of whose entries are 1, and **c** the $n \times 1$ matrix all of whose entries are 1. Compute **rc** and **cr**.

ex2.3.2 Prove Proposition 2.3.2.

ex2.3.3 A square matrix (a_{ij}) is *diagonal* if the entries off the diagonal are all zero, i.e., $i \neq j \implies a_{ij} = 0$.

---

[3]$\delta_{j,k}$ is the Kronecker delta, equal to 1 if $j = k$, and to 0 otherwise; $E_{i,j}$ is the matrix whose entries $a_{k,l}$ are zero except that $a_{i,j} = 1$.

Prove: If A is a diagonal matrix with distinct entries on the diagonal, and if B is a matrix such that $AB = BA$, then B is diagonal.

ex2.3.4 Denote by $\Xi(n; i, j)$, $1 \leq i \neq j \leq n$, the $n \times n$ matrix obtained from the identity by interchanging rows i and j, i.e.,

$$E_{ij} + E_{ji} + \sum_{k \neq ij} E_{kk}.$$

Let $A \in \mathcal{M}(n, m)$ and $B \in \mathcal{M}(m, n)$. Describe $\Xi(n; i, j)A$ and $B\Xi(n; i, j)$.

ex2.3.5 Let σ be a permutation of $[1, \ldots, n]$. Let A_σ be the $n \times n$ matrix whose entries a_{ij} are defined by

$$(2.3.6) \qquad\qquad a_{ij} = \begin{cases} 1 & \text{if } i = \sigma(j), \\ 0 & \text{otherwise.} \end{cases}$$

Show that A_σ has precisely one 1 in each row and in each column, and all of its other entries are 0. Conversely, show that if A is a matrix with precisely one 1 in each row and in each column, and all of its other entries are 0, then $A = A_\sigma$ for some permutation σ of $[1, \ldots, n]$. Such matrices are called *permutation matrices*.

What is the transpose $(A_\sigma)^{Tr}$ of the permutation matrix A_σ?

ex2.3.6 Show that the map $\sigma \mapsto A_\sigma$ defined above is multiplicative, that is: $A_{\sigma\tau} = A_\sigma A_\tau$ ($\sigma\tau$ is defined by composition: $\sigma\tau(j) = \sigma(\tau(j))$ for all $j \in [1, n]$).

ex2.3.7 Show that every permutation matrix $A_\sigma \in \mathcal{M}(n)$ is a product of matrices $\Xi(n; i, j)$.

Hint: See 4.1.2.

ex2.3.8 Let $A_\sigma \in \mathcal{M}(n)$ be a permutation matrix. Let $B \in \mathcal{M}(n, m)$ and $C \in \mathcal{M}(m, n)$. Describe $A_\sigma B$ and $C A_\sigma$.

ex2.3.9 Let $B \in \mathcal{M}(n)$ be skew-symmetric, $\sigma \in S_n$. Prove that $A_\sigma B A_\sigma^{-1}$ is skew-symmetric.

Similarly, if B is symmetric, then so is $A_\sigma B A_\sigma^{-1}$.

ex2.3.10 Denote by E_{ij}, $1 \leq i, j \leq n$, the $n \times n$ matrix whose entries are all zero except for the ij entry which is 1. Let $A \in \mathcal{M}(n, m)$ and $B \in \mathcal{M}(m, n)$. Describe $E_{ij}A$ and BE_{ij}.

ex2.3.11 Describe an $n \times n$ matrix $K(i, c)$ such that if $B \in \mathcal{M}(n, m)$, then $K(i, c)B$ is the matrix obtained from B by multiplying its i'th row by c.

ex2.3.12 Let $B \in \mathcal{M}(n,m)$. Describe a square matrix $A(c,i,j)$ such that multiplying B by it on the appropriate side has the effect of replacing the i'th row in B by the sum of the i'th row and c times the j'th row. Do the same for columns.

ex2.3.13 Show that each of the steps in the reduction of a matrix A to its row-echelon form (see 1.4.4) can be accomplished by left-multiplication of A by an appropriate matrix, so that the entire reduction to row-echelon form can be accomplished by left-multiplication by an appropriate matrix. Conclude that if the row rank of $A \in \mathcal{M}(n)$ is n, then A is left-invertible.

ex2.3.14 Let $A \in \mathcal{M}(n)$ be non-singular and let $B = (A,I)$, the matrix obtained by "augmenting" A by the identity matrix, that is, by adding to A the columns of I in their given order as columns $n+1, \dots, 2n$. Show that the matrix obtained by reducing B to row-echelon form is (I, A^{-1}).

ex2.3.15 Prove that if $A \in \mathcal{M}(n,m)$ and $B \in \mathcal{M}(m,l)$, then[4] $(AB)^{Tr} = B^{Tr}A^{Tr}$. Show that if $A \in \mathcal{M}(n)$ has a left inverse, then A^{Tr} has a right inverse, and if A has a right inverse, then A^{Tr} has a left inverse. Use the fact that A and A^{Tr} have the same rank to show that if A has a left inverse, B, it also has a right inverse, C, and since $B = B(AC) = (BA)C = C$, we have $BA = AB = I$ and A has an inverse.

Where does the fact that we deal with finite-dimensional spaces enter the proof?

ex2.3.16 Prove that the operator $P(x) \mapsto xP(x)$ on $\mathbb{F}[x]$ is left-invertible, but not right-invertible.

ex2.3.17 What are the ranks and the inverses (when they exist) of the matrices

$$
\begin{bmatrix} 0 & 2 & 1 & 0 \\ 1 & 1 & 7 & 1 \\ 2 & 2 & 2 & 2 \\ 0 & 5 & 0 & 0 \end{bmatrix},
\quad
\begin{bmatrix} 1 & 1 & 1 & 1 & 1 \\ 0 & 2 & 2 & 1 & 1 \\ 2 & 1 & 2 & 1 & 2 \\ 0 & 5 & 0 & 9 & 1 \\ 0 & 5 & 0 & 0 & 7 \end{bmatrix},
\quad
\begin{bmatrix} 1 & 1 & 1 & 1 & 1 \\ 0 & 1 & 1 & 1 & 1 \\ 0 & 0 & 1 & 1 & 1 \\ 0 & 0 & 0 & 1 & 1 \\ 0 & 0 & 0 & 0 & 1 \end{bmatrix}.
$$

ex2.3.18 Denote $A_n = \begin{bmatrix} 1 & n \\ 0 & 1 \end{bmatrix}$. Prove that $A_m A_n = A_{m+n}$ for all integers m, n.

---

[4]For the notation, see 2.1, example c.

ex2.3.19 Let $A, B, C, D \in \mathcal{M}(n)$, and let $\mathcal{E} = \begin{bmatrix} A & B \\ C & D \end{bmatrix} \in \mathcal{M}(2n)$ be the matrix whose top left quarter is a copy of A, the top right quarter a copy of B, etc. Prove that $\mathcal{E}^2 = \begin{bmatrix} A^2 + BC & AB + BD \\ CA + DC & CB + D^2 \end{bmatrix}$.

2.4   Matrices and operators

2.4.1   We have seen (2.1.7) that every matrix $A \in \mathcal{M}(m, n, \mathbb{F})$ defines an operator $T_A \in \mathcal{L}(\mathbb{F}^n_{\mathbf{c}}, \mathbb{F}^m_{\mathbf{c}})$. A quick check shows that T_A is the operator of left-multiplication by A of the columns of $\mathbb{F}^n_{\mathbf{c}}$. Conversely, given $T \in \mathcal{L}(\mathbb{F}^n_{\mathbf{c}}, \mathbb{F}^m_{\mathbf{c}})$, if we take $A = A_T$ to be the $m \times n$ matrix whose columns are $T e_j$, where $\{e_1, \dots, e_n\}$ is the standard basis in $\mathbb{F}^n$, we have $T_A = T$. In other words, the maps $A \mapsto T_A$ and $T \mapsto A_T$ are each the inverse of the other. Since both maps are clearly linear, we obtain the following theorem.

Theorem. *The map $T \mapsto A_T$ is an isomorphism of $\mathcal{L}(\mathbb{F}^n_{\mathbf{c}}, \mathbb{F}^m_{\mathbf{c}})$ onto $\mathcal{M}(m, n)$.*

2.4.2   Let $T \in \mathcal{L}(\mathbb{F}^n, \mathbb{F}^m)$, $S \in \mathcal{L}(\mathbb{F}^m, \mathbb{F}^l)$, and let $A_T \in \mathcal{M}(m, n)$, resp. $A_S \in \mathcal{M}(l, m)$ be the corresponding matrices; then

$$ST \in \mathcal{L}(\mathbb{F}^n, \mathbb{F}^l), \qquad A_S A_T \in \mathcal{M}(l, n),$$

and, since the multiplication in both sides is composition,

(2.4.1) $$A_{ST} = A_S A_T.$$

In particular, if $n = m = l$, we obtain

Theorem. *The map $T \mapsto A_T$ is an algebra isomorphism of $\mathcal{L}(\mathbb{F}^n)$ onto $\mathcal{M}(n)$.*

2.4.3   The special thing about $\mathbb{F}^n_{\mathbf{c}}$ and $\mathbb{F}^m_{\mathbf{c}}$ is that they have "standard bases". The correspondence $T \leftrightarrow A_T$ (or $A \leftrightarrow T_A$) uses these bases implicitly.

Consider now general finite-dimensional vector spaces $\mathcal{V}$ and $\mathcal{W}$. Let $T \in \mathcal{L}(\mathcal{V}, \mathcal{W})$, and let $\mathbf{v} = \{v_1, \dots, v_n\}$ be a basis for $\mathcal{V}$ and $\mathbf{w}$ a basis for $\mathcal{W}$.

Define the scalars t_{kj} by the expansion: $Tv_j = \sum_{k=1}^m t_{kj} w_k$, then, for any vector $v = \sum c_j v_j$, we have

$$(2.4.2) \qquad Tv = \sum c_j Tv_j = \sum_j \sum_k c_j t_{k,j} w_k = \sum_k \left(\sum_j c_j t_{k,j} \right) w_k.$$

Given the bases $\{v_1, \ldots, v_n\}$ and $\{w_1, \ldots, w_m\}$, the full information about T is contained in the matrix

$$(2.4.3) \qquad A_{T,\mathbf{v},\mathbf{w}} = \begin{bmatrix} t_{11} & \cdots & t_{1n} \\ t_{21} & \cdots & t_{2n} \\ \vdots & \cdots & \vdots \\ t_{m1} & \cdots & t_{mn} \end{bmatrix} = [C_{\mathbf{w}} Tv_1, \ldots, C_{\mathbf{w}} Tv_n].$$

The *"coordinates operators"*, $C_{\mathbf{w}}$, assign to each vector in $\mathscr{W}$ the column of its coordinates with respect to the basis $\mathbf{w}$; see (2.1.8).

Given the bases $\mathbf{v}$ and $\mathbf{w}$, and the matrix $A_{T,\mathbf{v},\mathbf{w}}$, the operator T is explicitly defined by (2.4.2), which says: *The column of $\mathbf{w}$-coordinates of Tv is obtained by multiplying the column of $\mathbf{v}$-coordinates of v on the left by the matrix $A_{T,\mathbf{v},\mathbf{w}}$; that is:*

$$(2.4.4) \qquad C_{\mathbf{w}} Tv = A_{T,\mathbf{v},\mathbf{w}} C_{\mathbf{v}} v.$$

Let $A \in \mathscr{M}(m,n)$, and denote by Sv the vector in $\mathscr{W}$ whose coordinates with respect to $\mathbf{w}$ are given by the column $A C_{\mathbf{v}} v$. So defined, S is clearly a linear operator in $\mathscr{L}(\mathscr{V}, \mathscr{W})$ and $A_{S,\mathbf{v},\mathbf{w}} = A$. This gives:

Theorem. *Given vector spaces $\mathscr{V}$ and $\mathscr{W}$ with bases $\mathbf{v} = \{v_1, \ldots, v_n\}$ and $\mathbf{w} = \{w_1, \ldots, w_m\}$ repectively, the map $T \mapsto A_{T,\mathbf{v},\mathbf{w}}$ is an isomorphism of $\mathscr{L}(\mathscr{V}, \mathscr{W})$ onto $\mathscr{M}(m,n)$.*

When $\mathscr{W} = \mathscr{V}$ and $\mathbf{w} = \mathbf{v}$, we write $A_{T,\mathbf{v}}$ instead of $A_{T,\mathbf{v},\mathbf{v}}$. The reader should check that, just as in subsection 2.4.2, the isomorphism $T \mapsto A_{T,\mathbf{v}}$ is *an algebra isomorphism* of $\mathscr{L}(\mathscr{V})$ onto $\mathscr{M}(n)$.

2.4.4 Change of basis. Assume now that $\mathscr{W} = \mathscr{V}$, and that $\mathbf{v}$ and $\mathbf{w}$ are arbitrary bases. The $\mathbf{v}$-coordinates of a vector v are given by $C_{\mathbf{v}} v$, and the $\mathbf{w}$-coordinates of v by $C_{\mathbf{w}} v$. If we are given the $\mathbf{v}$-coordinates of a vector v, say $x = C_v v$, and we need the $\mathbf{w}$-coordinates of v, we observe that $v = C_v^{-1} x$, and hence $C_\mathbf{w} v = C_\mathbf{w} C_v^{-1} x$. In other words, the operator

$$(2.4.5) \qquad\qquad C_{\mathbf{w},\mathbf{v}} = C_\mathbf{w} C_\mathbf{v}^{-1}$$

on $\mathbb{F}^n$ assigns to the (column of) $\mathbf{v}$-coordinates of a vector $v \in \mathcal{V}$ the column of its $\mathbf{w}$-coordinates. The opeator $C_\mathbf{v}^{-1}$ identifies the vector from its $\mathbf{v}$-coordinates, and $C_\mathbf{w}$ assigns to the identified vector its $\mathbf{w}$-coordinates; the space $\mathcal{V}$ remains in the background. Notice that $C_{\mathbf{v},\mathbf{w}}^{-1} = C_{\mathbf{w},\mathbf{v}}$.

Suppose that we have the matrix $A_{T,\mathbf{w}}$ of an operator $T \in \mathscr{L}(\mathcal{V})$ relative to a basis $\mathbf{w}$, and we want to have the matrix $A_{T,\mathbf{v}}$ of the same operator T, but relative to a basis $\mathbf{v}$. Claim:

$$(2.4.6) \qquad\qquad A_{T,\mathbf{v}} = C_{\mathbf{v},\mathbf{w}} A_{T,\mathbf{w}} C_{\mathbf{w},\mathbf{v}};$$

$C_{\mathbf{w},\mathbf{v}}$ assigns to the $\mathbf{v}$-coordinates of a vector $v \in \mathcal{V}$ its $\mathbf{w}$-coordinates; $A_{T,\mathbf{w}}$ replaces the $\mathbf{w}$-coordinates of v by those of Tv; $C_{\mathbf{v},\mathbf{w}}$ identifies Tv from its $\mathbf{w}$-coordinates, and produces its $\mathbf{v}$-coordinates.

2.4.5   How special are the matrices (operators) $C_{\mathbf{w},\mathbf{v}}$? They are clearly invertible, and that is a complete characterization.

Proposition. *Given a basis $\mathbf{w} = \{w_1, \ldots, w_n\}$ of $\mathcal{V}$, the map $\mathbf{v} \mapsto C_{\mathbf{w},\mathbf{v}}$ is a bijection of the set of bases $\mathbf{v}$ of $\mathcal{V}$ onto $\mathbf{GL}(n, \mathbb{F})$.*

PROOF: Since $C_\mathbf{w}$ is non-singular, the equality $C_{\mathbf{w},\mathbf{v}_1} = C_{\mathbf{w},\mathbf{v}_2}$ implies $C_{\mathbf{v}_1}^{-1} = C_{\mathbf{v}_2}^{-1}$, and since $C_{\mathbf{v}_1}^{-1}$ maps the elements of the standard basis of $\mathbb{F}^n$ onto the corresponding elements in $\mathbf{v}_1$, and $C_{\mathbf{v}_2}^{-1}$ maps the same vectors onto the corresponding elements in $\mathbf{v}_2$, we have $\mathbf{v}_1 = \mathbf{v}_2$. This proves the injectivity.

To prove the surjectivity, let $S \in \mathbf{GL}(n, \mathbb{F})$ be arbitrary. We shall exhibit a base $\mathbf{v}$ such that $S = C_{\mathbf{w},\mathbf{v}}$. By definition, $C_\mathbf{w} w_j = e_j$ (recall that $\{e_1, \ldots, e_n\}$ is the standard basis for $\mathbb{F}^n$). Define the vectors v_j by the condition: $C_\mathbf{w} v_j = S e_j$, that is, v_j is the vector whose $\mathbf{w}$-coordinates are given by the j'th column of the matrix S. As S is non-singular the v_j's are linearly independent, hence form a basis $\mathbf{v}$ of $\mathcal{V}$.

For all j we have $v_j = C_\mathbf{v}^{-1} e_j$ and $C_{\mathbf{w},\mathbf{v}} e_j = C_\mathbf{w} v_j = S e_j$. This proves that $S = C_{\mathbf{w},\mathbf{v}}$. ◀

2.4.6 Similarity (matrices). By definition, the matrices B_1 and B_2 are *similar* if they represent the same operator T in terms of (possibly) different bases, that is, $B_1 = A_{T,\mathbf{v}}$ and $B_2 = A_{T,\mathbf{w}}$.

If B_1 and B_2 are similar, they are related by (2.4.6). By Proposition 2.4.5 we have

Proposition. *The matrices B_1 and B_2 are similar if and only if there exists $C \in GL(n,\mathbb{F})$ such that*

$$(2.4.7) \qquad B_1 = C^{-1} B_2 C, \quad \text{or equivalently,} \quad CB_1 = B_2 C.$$

In other words, *similarity* for A and B is synonymous to *conjugation* (under $GL(n,\mathbb{F})$). We shall see later (see exercises **ex4.5.18** and **ex7.5.4**) that if there exists such C for (2.4.7) with entries in some field extension of $\mathbb{F}$, then there exists one in $\mathscr{M}(n,\mathbb{F})$.

A matrix is *diagonalizable* if it is similar to a diagonal matrix.

2.4.7 Similarity (operators). The operators $S, T \in \mathscr{L}(\mathscr{V})$ are said to be *similar* if they are conjugate under $GL(\mathscr{V})$, that is, if there is an operator $R \in GL(\mathscr{V})$ such that

$$(2.4.8) \qquad T = RSR^{-1} \quad \text{or, equivalently,} \quad RS = TR.$$

An operator is *diagonalizable* if its matrix (relative to some basis) is diagonalizable[5]. Observe that the matrix $A_{T,\mathbf{v}}$ of T relative to a basis $\mathbf{v} = \{v_1, \ldots, v_n\}$ is diagonal if and only if $Tv_j = \lambda_j v_j$, where λ_j is the j'th entry on the diagonal of $A_{T,\mathbf{v}}$.

2.4.8 Similarity (linear systems). A *linear system* is a pair $(\mathscr{V}, T)$ where $\mathscr{V}$ is a vector space and $T \in \mathscr{L}(\mathscr{V})$.

---

[5]Because of Theorem 2.4.2, we apply to matrices the language and terminology introduced for operators, and vice versa.

DEFINITION: The systems $(\mathscr{V}_1, T_1)$ and $(\mathscr{V}_2, T_2)$ are *similar* (other terms used: *conjugate*, and *isomorphic*) if there is an isomorphism Ψ of $\mathscr{V}_1$ onto $\mathscr{V}_2$ such that

(2.4.9) $$\Psi T_1 = T_2 \Psi.$$

The condition (2.4.9) is often described by stating that the following is a *commutative diagram:*

(2.4.10)

$$
\begin{array}{ccc}
\mathscr{V}_1 & \xrightarrow{\;T_1\;} & \mathscr{V}_1 \\
\downarrow{\scriptstyle \Psi} & & \downarrow{\scriptstyle \Psi} \\
\mathscr{V}_2 & \xrightarrow{\;T_2\;} & \mathscr{V}_2
\end{array}
$$

EXAMPLE: *Right-multiplication* by matrices in $\mathscr{M}(n)$ defines linear operators on $\mathbb{F}_{\mathbf{r}}^n$. That is, $A \in \mathscr{M}(n)$ defines the linear map R_A by:

(2.4.11) $$R_A [a_1, \ldots, a_n] = [a_1, \ldots, a_n] A.$$

If we take the transpose on both sides, we obtain

(2.4.12) $$[R_A [a_1, \ldots, a_n]]^{Tr} = A^{Tr} [a_1, \ldots, a_n]^{Tr}.$$

Let Ψ denote the isomorphism of $\mathbb{F}_{\mathbf{r}}^n$ onto $\mathbb{F}_{\mathbf{c}}^n$ given by transposition. The equality (2.4.12) is identical with (2.4.9) where $T_1 = R_A$, and T_2 is left-multiplication by A^{Tr} on $\mathbb{F}_{\mathbf{c}}^n$. Thus, right-multiplication by A on $\mathbb{F}_{\mathbf{r}}^n$ is similar to left-multiplication by A^{Tr} on $\mathbb{F}_{\mathbf{c}}^n$.

EXERCISES FOR SECTION 2.4

ex2.4.1 Prove that $S, T \in \mathscr{L}(\mathscr{V})$ are similar if and only if their matrices (relative to any basis) are similar. An equivalent condition is: for any basis $\mathbf{w}$ there is a basis $\mathbf{v}$ such that $A_{T,\mathbf{v}} = A_{S,\mathbf{w}}$.

ex2.4.2 Prove that two linear systems, $(\mathscr{V}, T)$ and $(\mathscr{W}, S)$, are *similar* if and only if $\dim \mathscr{V} = \dim \mathscr{W}$ and, if $\mathbf{v}$ is a basis for $\mathscr{V}$ and $\mathbf{w}$ a basis for $\mathscr{W}$, then $A_{T,\mathbf{v}}$ and $A_{S,\mathbf{w}}$, the matrices of T and S relative to these bases, are similar.

ex2.4.3 Let R_A be defined on $\mathbb{F}_{\mathbf{r}}^n$ by (2.4.11).

a. What are the images of the elements of the standard basis in $\mathbb{F}_{\mathbf{r}}^n$?

b. Prove that the correspondence $A \leftrightarrow R_A$ is an isomorphism of vector spaces. Is it an algebra isomorphism?

ex2.4.4 Let $\mathbb{F}_n[x]$ be the space of polynomials $\sum_0^n a_j x^j$. Let D be the differentiation operator and $T = 2D + I$.

a. What is the matrix corresponding to T relative to the basis $\{x^j\}_{j=0}^n$?

b. Verify that, if $u_j = \sum_{l=j}^n x^l$, then $\{u_j\}_{j=0}^n$ is a basis, and find the matrix corresponding to T relative to this basis.

ex2.4.5 Let $\mathscr{T}_N$ be the complex vector space of 2π-periodic trigonometric polynomials of degree $\leq N$; that is, the space of functions f of the form $f(t) = \sum_{|n| \leq N} a_n e^{int}$. Let $D : f \mapsto f'$ be the differentiation operator, and $T = D^2 + 2D$. What is the matrix of T relative to the basis $\{e^{int}\}_{|n| \leq N}$?

ex2.4.6 Let $\mathscr{V}$ be n-dimensional, $v \in \mathscr{V}$, and $T \in \mathscr{L}(\mathscr{V})$ such that $\{T^j v\}_{j=0}^{n-1}$ are linearly independent.

a. Verify: $\{T^j v\}_{j=0}^{n-1}$ is a basis for $\mathscr{V}$, and hence $T^n v = \sum_{j=0}^{n-1} a_j T^j v$ (for appropriate coefficients a_j).

b. What is the matrix of T relative to the basis $\{T^j v\}_{j=0}^{n-1}$?

ex2.4.7 Let $\mathscr{V}$ be n-dimensional, $v, w \in \mathscr{V}$, and $T \in \mathscr{L}(\mathscr{V})$.

a. Assume that the subspace $\mathscr{V}_1$ spanned by $\{T^j v\}_{j=0}^{n-1}$ is k-dimensional. Prove that $\{T^j v\}_{j=0}^{k-1}$ is linearly independent (and hence is a basis for $\mathscr{V}_1$).

b. With v as above, assume that $\mathscr{V}_2$, the span of $\{T^j v\}_{j=0}^{n-1} \cup \{T^j w\}_{j=0}^{n-1}$, is m-dimensional. Prove that $\{T^j v\}_{j=0}^{k-1} \cup \{T^j w\}_{j=0}^{m-k-1}$ is linearly independent (and hence is a basis for $\mathscr{V}_2$).

ex2.4.8 Let $A \in \mathscr{M}(l,m)$. Prove that the map $T : B \mapsto AB$ of $\mathscr{M}(m,n)$ into $\mathscr{M}(l,n)$ is a linear operator. In particular, if $n = 1$, $\mathscr{M}(m,1) = \mathbb{F}_{\mathbf{c}}^m$, $\mathscr{M}(l,1) = \mathbb{F}_{\mathbf{c}}^l$ and $T \in \mathscr{L}(\mathbb{F}_{\mathbf{c}}^m, \mathbb{F}_{\mathbf{c}}^l)$. What is the relation between A and the matrix A_T defined in 2.4.3 (for the standard bases, and with n there replaced here by l)?

2.5 Kernel, range, nullity, and rank

2.5.1 DEFINITION: The *kernel* of an operator $T \in \mathscr{L}(\mathscr{V}, \mathscr{W})$ is the set

$$\ker(T) = \{v \in \mathscr{V} : Tv = 0\}.$$

The *range* of T is the set

$$\text{range}(T) = T\mathscr{V} = \{w \in \mathscr{W} : w = Tv \ \text{ for some } v \in \mathscr{V}\}.$$

The kernel is also called the *nullspace* of T. An operator T is called *nonsingular* if $\ker(T) = \{0\}$ and *singular* otherwise.

Proposition. *Assume* $T \in \mathscr{L}(\mathscr{V}, \mathscr{W})$. *Then* $\ker(T)$ *is a subspace of* $\mathscr{V}$, *and* $\text{range}(T)$ *is a subspace of* $\mathscr{W}$.

PROOF: If $v_1, v_2 \in \ker(T)$, then $T(a_1 v_1 + a_2 v_2) = a_1 T v_1 + a_2 T v_2 = 0$. To show that $\text{range}(T)$ is a subspace of $\mathscr{W}$ observe that if $w_j = T v_j$, $j = 1, 2$, then $a_1 w_1 + a_2 w_2 = T(a_1 v_1 + a_2 v_2)$.   ◄

If $\mathscr{V}$ is finite-dimensional and $T \in \mathscr{L}(\mathscr{V}, \mathscr{W})$ then both $\ker(T)$ and $\text{range}(T)$ are finite-dimensional; the first since it is a subspace of a finite-dimensional space, and the second as the image of one (since, if $\{v_1, \ldots, v_n\}$ is a basis for $\mathscr{V}$, $\{T v_1, \ldots, T v_n\}$ spans $\text{range}(T)$).

We define $\rho(T)$, the *rank* of T, as the dimension of $\text{range}(T)$. We define $\nu(T)$, the *nullity* of T, as the dimension of $\ker(T)$.

Theorem (Rank and nullity). *Let* $\mathscr{V}$ *be finite-dimensional,* $\mathscr{W}$ *arbitrary, and* $T \in \mathscr{L}(\mathscr{V}, \mathscr{W})$. *Then*

$$(2.5.1) \qquad\qquad \rho(T) + \nu(T) = \dim \mathscr{V}.$$

PROOF: Let $\{v_1, \ldots, v_l\}$ be a basis for $\ker(T)$, $l = \nu(T)$, and extend it to a basis of $\mathscr{V}$ by adding $\{u_1, \ldots, u_k\}$. By 1.3.5 we have $l + k = \dim \mathscr{V}$. The theorem follows if we show that $k = \rho(T)$. We do this by showing that $\{T u_1, \ldots, T u_k\}$ is a basis for $\text{range}(T)$.

Write any $v \in \mathscr{V}$ as $\sum_{i=1}^{l} a_i v_i + \sum_{i=1}^{k} b_i u_i$. Then $Tv = \sum_{i=1}^{k} b_i T u_i$ (since $T v_i = 0$). This shows that $\{T u_1, \ldots, T u_k\}$ spans $\text{range}(T)$.

To prove that $\{T u_1, \ldots, T u_k\}$ is also independent, we observe that if $\sum_{j=1}^{k} c_j T u_j = 0$, then $T\left(\sum_{j=1}^{k} c_j u_j\right) = 0$, that is, $\sum_{j=1}^{k} c_j u_j \in \ker(T)$. Since $\{v_1, \ldots, v_l\}$ is a basis for $\ker(T)$, we have $\sum_{j=1}^{k} c_j u_j = \sum_{j=1}^{l} d_j v_j$ for appropriate constants d_j. But $\{v_1, \ldots, v_l\} \cup \{u_1, \ldots, u_k\}$ is independent, and it follows that $c_j = 0$ (and $d_j = 0$) for all j.   ◄

The proof gives more than is claimed in the theorem. It shows that T can be "factored" as a product of two maps:

(2.5.2) $\qquad \mathscr{V} \to \mathscr{V}/\ker(T), \qquad$ and $\qquad \mathscr{V}/\ker(T) \to T\mathscr{V}.$

The first is the quotient map $\mathscr{V} \to \mathscr{V}/\ker(T)$; vectors that are congruent modulo $\ker(T)$ have the same image under T, which means that T assigns the same image to the elements of each coset, and different images to elements of distinct cosets. Assigning to cosets the common image of its elements defines the second map, $\mathscr{V}/\ker(T) \to TV$ which is both injective and surjective, i.e., an isomorphism. (This is the *Homomorphism Theorem* of groups in our context.)

2.5.2 The identity operator, defined by $Iv = v$, is an identity element in the algebra $\mathscr{L}(\mathscr{V})$. The invertible elements in $\mathscr{L}(\mathscr{V})$ are the *automorphisms* of $\mathscr{V}$, that is, the *bijective* linear maps.

In the context of operators on finite-dimensional spaces, and more generally, between vector spaces of the same finite dimension, *injectivity* and *surjectivity* are equivalent properties.

Theorem. *Let $\mathscr{V}$ be a finite-dimensional vector space, $T \in \mathscr{L}(\mathscr{V})$. Then*

(2.5.3) $\qquad\qquad \ker(T) = \{0\} \iff \mathrm{range}(T) = \mathscr{V},$

and both conditions are equivalent to T being invertible.

PROOF: $\ker(T) = \{0\}$ is equivalent to $v(T) = 0$, and $\mathrm{range}(T) = \mathscr{V}$ is equivalent to $\rho(T) = \dim \mathscr{V}$. Now apply (2.5.1). ◀

It follows that for an operator T on a finite-dimensional space V, invertibility and nonsingularity are equivalent.

2.5.3 As another illustration of how the "rank and nullity" theorem can be used, consider the following statement (which can be seen directly as a consequence of exercise **ex1.3.10**).

Theorem. *Let $\mathscr{V} = \mathscr{V}_1 \oplus \mathscr{V}_2$ be finite-dimensional, and $\dim \mathscr{V}_1 = k$. Let $\mathscr{W} \subset \mathscr{V}$ be a subspace of dimension $l > k$. Then $\dim \mathscr{W} \cap \mathscr{V}_2 \geq l - k$.*

PROOF: Denote by π_1 the restriction to $\mathcal{W}$ of the projection of $\mathcal{V}$ on $\mathcal{V}_1$ along $\mathcal{V}_2$. Since the rank of π_1 is clearly $\leq k$, its nullity is $\geq l - k$. In other words, the dimension of the kernel of this map, $\dim \mathcal{W} \cap \mathcal{V}_2$, is at least $l - k$.   ◀

EXERCISES FOR SECTION 2.5

ex2.5.1 Assume that $\mathcal{V}$ is finite-dimensional and $T \in \mathcal{L}(\mathcal{V},\mathcal{W})$. Let **v** and **w** be bases for $\mathcal{V}$ and $\mathcal{W}$ respectively. Prove that $\rho(T)$ is equal to the rank $\rho(A_{T,\mathbf{v},\mathbf{w}})$ of the matrix $A_{T,\mathbf{v},\mathbf{w}}$.

ex2.5.2 Assume $T, S \in \mathcal{L}(\mathcal{V})$. Prove that $\nu(ST) \leq \nu(S) + \nu(T)$.

ex2.5.3 Let $T \in \mathcal{L}(\mathcal{V})$, and let $\mathcal{W} \subset \mathcal{V}$ be a subspace such that $\mathcal{W} \supset \ker(T)$, and $T\mathcal{W} = T\mathcal{V}$. Prove that $\mathcal{W} = \mathcal{V}$.

ex2.5.4 Give an example of two 2×2 matrices A and B such that $\rho(AB) = 1$ and $\rho(BA) = 0$.

ex2.5.5 Assume $\dim \mathcal{V} = n$, $T, S \in \mathcal{L}(\mathcal{V})$. Prove

$$(2.5.4) \qquad\qquad \rho(TS) = \rho(S) - \dim(S\mathcal{V} \cap \ker(T)).$$

ex2.5.6 Given vector spaces $\mathcal{V}$ and $\mathcal{W}$ over the same field. Let $\{v_j\}_{j=1}^n \subset \mathcal{V}$ and $\{w_j\}_{j=1}^n \subset \mathcal{W}$. Prove that there exists a linear map T of $\mathrm{span}[v_1, \ldots, v_n]$ into $\mathcal{W}$ such that $Tv_j = w_j$ for all j if and only if the following implication holds:

$$\text{If } a_j, \ j = 1, \ldots, n, \ \text{are scalars, and } \ \sum_1^n a_j v_j = 0, \quad \text{then} \quad \sum_1^n a_j w_j = 0.$$

Can the definition of T be extended to the entire $\mathcal{V}$?

ex2.5.7 What is the relationship of the previous exercise to Theorem 1.4.5?

ex2.5.8 The operators $T, S \in \mathcal{L}(\mathcal{V})$ are called *"equivalent"* if there exist invertible $A, B \in \mathcal{L}(\mathcal{V})$ such that

$$S = ATB \quad \text{(so that } T = A^{-1}SB^{-1}).$$

Prove that if $\mathcal{V}$ is finite-dimensional, then T, S are *"equivalent"* if and only if

$$\rho(S) = \rho(T).$$

ex2.5.9 Give an example of two operators on $\mathbb{R}^3$ that are "equivalent" but not similar.

ex2.5.10 Assume $T, S \in \mathcal{L}(\mathcal{V})$. Prove that the following statements are equivalent:

a. $\ker(S) \subset \ker(T)$;

b. There exists $R \in \mathcal{L}(\mathcal{V})$ such that $T = RS$.

Hint: For the implication **a** $\Longrightarrow$ **b**: choose a basis $\{v_1, \dots, v_s\}$ for $\ker(S)$. Expand it to a basis for $\ker(T)$ by adding $\{u_1, \dots, u_{t-s}\}$, and expand further to a basis for $\mathcal{V}$ by adding the vectors $\{w_1, \dots, w_{n-t}\}$.

The sequence $\{Su_1, \dots, Su_{t-s}\} \cup \{Sw_1, \dots, Sw_{n-t}\}$ is independent, so that R can be defined arbitrarily on it (and extended by linearity to an operator on the entire space). Define $R(Su_j) = 0$, $R(Sw_j) = Tw_j$.

The other implication is obvious.

ex2.5.11 Assume $T, S \in \mathcal{L}(\mathcal{V})$. Prove that the following statements are equivalent:

a. $\mathrm{range}(S) \subset \mathrm{range}(T)$;

b. There exists $R \in \mathcal{L}(\mathcal{V})$ such that $S = TR$.

Hint: Again, **b** $\Longrightarrow$ **a** is obvious.

For **a** $\Longrightarrow$ **b** take a basis $\{v_1, \dots, v_n\}$ for $\mathcal{V}$. Let u_j, $j = 1, \dots, n$, be such that $Tu_j = Sv_j$ (use assumption **a**). Define $Rv_j = u_j$ (and extend by linearity).

ex2.5.12 Find bases for the nullspace, $\ker(A)$, and for the range, $\mathrm{range}(A)$, of the matrix, acting by right-multiplication on (rows in) $\mathbb{R}_r^5$,

$$
\begin{bmatrix}
1 & 0 & 0 & 5 & 9 \\
0 & 1 & 0 & -3 & 2 \\
0 & 0 & 1 & 2 & 1 \\
3 & 2 & 1 & 11 & 32 \\
1 & 2 & 0 & -1 & 13
\end{bmatrix}.
$$

ex2.5.13 Let $T \in \mathcal{L}(V)$, $l \in \mathbb{N}$. Prove:

a. $\ker(T^l) \subseteq \ker(T^{l+1})$; equality if and only if $\mathrm{range}(T^l) \cap \ker(T) = \{0\}$.

b. $\mathrm{range}(T^{l+1}) \subseteq \mathrm{range}(T^l)$; equality if and only if $\ker(T^{l+1}) = \ker(T^l)$.

c. If $\ker(T^{l+1}) = \ker(T^l)$, then $\ker(T^{l+k+1}) = \ker(T^{l+k})$ for every positive integer k.

*ex2.5.14 Prove that the rank of a skew-symmetric matrix is even. (See 1.2.3 example *d*.)

*2.6   Operator norms

2.6.1 If $\mathcal{V}$ and $\mathcal{W}$ are normed finite-dimensional vector spaces (see *1.5), we define a norm on $\mathcal{L}(\mathcal{V}, \mathcal{W})$ by writing, for $T \in \mathcal{L}(\mathcal{V}, \mathcal{W})$,

$$(2.6.1) \qquad \|T\| = \max_{\|v\|=1} \|Tv\| = \max_{v \neq 0} \frac{\|Tv\|}{\|v\|}.$$

Equivalently,

$$(2.6.2) \qquad \|T\| = \inf\{C : \|Tv\| \leq C\|v\| \text{ for all } v \in \mathcal{V}\}.$$

To check that (2.6.1) defines a norm, we observe that properties **a** and **b** (in 1.5.1) are obvious, and that **c** follows from

$$\|(T+S)v\| \leq \|Tv\| + \|Sv\| \leq \|T\|\|v\| + \|S\|\|v\| \leq (\|T\| + \|S\|)\|v\|.$$

The norms appearing in the inequalities are the ones defined on $\mathcal{W}$, $\mathcal{L}(\mathcal{V}, \mathcal{W})$, and $\mathcal{V}$, respectively.

The space $\mathcal{L}(\mathcal{V})$ is an algebra and we observe that the norm defined by (2.6.1) on $\mathcal{L}(\mathcal{V})$ is *submultiplicative*: given $S, T \in \mathcal{L}(\mathcal{V})$, then, for every $v \in \mathcal{V}$, $\|STv\| \leq \|S\|\|Tv\| \leq \|S\|\|T\|\|v\|$, which means

$$(2.6.3) \qquad \|ST\| \leq \|S\|\|T\|.$$

Remark: If $\mathcal{V}$ is real or complex, so is $\mathcal{L}(\mathcal{V})$ and, as remarked in 1.5.1, the notions of point-set topology are well defined, independently of the particular norm that we may be using.

EXERCISES FOR SECTION 2.6

ex2.6.1 Let $\mathcal{V}$ be a normed linear space and $T \in \mathcal{L}(\mathcal{V})$. Prove that the set of vectors $v \in \mathcal{V}$ whose T-orbit, $\{T^n v\}$, is bounded is a subspace of $\mathcal{V}$.

Chapter 3

Duality of Vector Spaces

3.1 Linear functionals

All the vector spaces we consider in this chapter are assumed to be finite-dimensional.

DEFINITION: A *linear functional*, a.k.a. *linear form*, on a vector space $\mathscr{V}$ is a linear map v^* of $\mathscr{V}$ into the underlying field $\mathbb{F}$.

As mentioned already in 2.1.3, the space $\mathscr{L}(\mathscr{V}, \mathbb{F})$ is called the *dual space* of $\mathscr{V}$, and is commonly denoted $\mathscr{V}^*$. As seen in 2.1.5, $\mathscr{V}^*$ with the standard operations (addition of operators and multiplication of operators by scalars) is a vector space over $\mathbb{F}$.

3.1.1 Linear functionals arise naturally in the expansion of vectors as linear combinations of elements of a basis.

Let $\{v_1, \ldots, v_n\}$ be a basis for $\mathscr{V}$. Every element $v \in \mathscr{V}$ can be written, in exactly one way, as

$$(3.1.1) \qquad v = \sum_1^n a_j(v) v_j.$$

The notation $a_j(v)$ makes explicit the dependence of the coefficients on the vector v.

Let $v = \sum_1^n a_j(v) v_j$ and $u = \sum_1^n a_j(u) v_j$. If $c, d \in \mathbb{F}$, then

$$cv + du = \sum_1^n (ca_j(v) + da_j(u)) v_j$$

so that

$$a_j(cv + du) = ca_j(v) + da_j(u).$$

In other words, the coefficient functions $a_j(v)$, $j = 1, \ldots, n$, are linear functionals on $\mathscr{V}$.

By Theorem 2.1.2, linear functionals are completely determined by the values they assign to the elements of any basis and, given a basis, one can assign arbitrary values to its elements and extend by linearity to a linear functional. The coefficient (or coordinate) functionals $a_j(v)$ of (3.1.1) can be defined in this manner, simply by the fact that $a_j(v)$ assigns the value 1 to v_j and 0 to v_k for $k \neq j$.

A standard notation for the image of a vector v under a linear functional v^* is (v, v^*). Accordingly we denote the linear functionals corresponding to $a_j(v)$ by v_j^* and write $a_j(v) = (v, v_j^*)$ so that

$$(3.1.2) \qquad\qquad v = \sum_j (v, v_j^*) v_j.$$

Proposition. *The linear functionals v_j^*, $j = 1, \ldots, n$, form a basis for the dual space $\mathcal{V}^*$.*

PROOF: SPANNING. Let $u^* \in \mathcal{V}^*$; write $b_j(u^*) = (v_j, u^*)$. We claim that $u^* = \sum b_j(u^*) v_j^*$.

By (3.1.2), if $v \in \mathcal{V}$,

$$(v, u^*) = \left(\sum_j (v, v_j^*) v_j,\, u^* \right) = \sum_j (v, v_j^*) b_j(u^*) = \left(v,\, \sum_j b_j(u^*) v_j^* \right),$$

and, since assigning the same value to every $v \in \mathcal{V}$ means equality for functionals, we have $u^* = \sum b_j(u^*) v_j^*$.

INDEPENDENCE. For any coefficients $c_j \in \mathbb{F}$, $j = 1, \ldots, n$, we have $c_k = (v_k, \sum c_j v_j^*)$, so that if $\sum c_j v_j^* = 0$, then $c_k = 0$ for all k. ◀

Corollary. $\dim \mathcal{V}^* = \dim \mathcal{V}$.

The basis $\{v_j^*\}_1^n$, $j = 1, \ldots, n$, is called the *dual basis* of $\{v_1, \ldots, v_n\}$. It is characterized by the condition

$$(3.1.3) \qquad\qquad (v_j, v_k^*) = \delta_{j,k} = \begin{cases} 1 & \text{if } j = k, \\ 0 & \text{otherwise.} \end{cases}$$

$\delta_{j,k}$ is known as the *Kronecker delta*.

3.1.2 The vector space structure defined on $\mathscr{V}^*$ guarantees that for every $v \in \mathscr{V}$ the map $v^* \mapsto (v, v^*)$ is a linear map from $\mathscr{V}^*$ to $\mathbb{F}$, that is, a linear functional on $\mathscr{V}^*$.

If $\{v_1, \ldots, v_n\}$ is a basis for $\mathscr{V}$, and $\{v_1^*, \ldots, v_n^*\}$ the dual basis in $\mathscr{V}^*$, then (3.1.3) identifies $\mathscr{V}$ as the dual of $\mathscr{V}^*$ and $\{v_1, \ldots, v_n\}$ as the dual basis of $\{v_1^*, \ldots, v_n^*\}$. This means that every linear functional on $\mathscr{V}^*$ is obtained as $v^* \mapsto (v, v^*)$ for some $v \in \mathscr{V}$. The roles of $\mathscr{V}$ and $\mathscr{V}^*$ are perfectly symmetric, and what we have is *two spaces in duality*, the duality between them defined by the *bilinear* (that is, linear in v for every fixed v^*, and linear in v^* for every fixed v) form (v, v^*). Expansion (3.1.2) works in both directions; thus if $\{v_1, \ldots, v_n\}$ and $\{v_1^*, \ldots, v_n^*\}$ are dual bases, then for all $v \in \mathscr{V}$ and $v^* \in \mathscr{V}^*$,

$$(3.1.4) \qquad v = \sum_1^n (v, v_j^*) v_j, \qquad v^* = \sum_1^n (v_j, v^*) v_j^*.$$

The dual of $\mathbb{F}_{\mathbf{c}}^n$ (i.e., $\mathbb{F}^n$ written as columns) can be identified with $\mathbb{F}_{\mathbf{r}}^n$ (i.e., $\mathbb{F}^n$ written as rows) and the pairing (v, v^*) as the matrix product $v^* v$ of the row v^* by the column v (exercise **ex3.1.5** below). The dual of the standard basis of $\mathbb{F}_{\mathbf{c}}^n$ is the standard basis $\mathbb{F}_{\mathbf{r}}^n$.

3.1.3 Annihilator. Given a set $A \subset \mathscr{V}$, the *annihilator* of A is the set $A^\perp$ of all the linear functionals $v^* \in \mathscr{V}^*$ that vanish identically on A. Formally:

$$(3.1.5) \qquad A^\perp = \{v^* \in \mathscr{V}^* : \ \forall v \in A, \ (v, v^*) = 0.\}$$

Clearly, $A^\perp$ is a subspace of $\mathscr{V}^*$.

Linear functionals that annihilate A vanish on $\mathrm{span}[A]$ as well, and functionals that annihilate $\mathrm{span}[A]$ clearly vanish on its subset A; hence

$$A^\perp = (\mathrm{span}[A])^\perp.$$

In particular, if $\{v_1, \ldots, v_m\}$ is a basis for a subspace $\mathscr{V}_1 \subset \mathscr{V}$, then $v^* \in \mathscr{V}_1^\perp$ if and only if $(v_j, v^*) = 0$ for $j = 1, \ldots, m$.

Theorem. *Let $\mathscr{V}_1 \subset \mathscr{V}$ be a subspace; then* $\dim \mathscr{V}_1 + \dim \mathscr{V}_1^\perp = \dim \mathscr{V}$.

PROOF: Let $\{v_1, \ldots, v_m\}$ be a basis for $\mathcal{V}_1$, and let $\{v_{m+1}, \ldots, v_n\}$ complete it to a basis for $\mathcal{V}$. Let $\{v_1^*, \ldots, v_n^*\}$ be the dual basis.

We claim that $\{v_{m+1}^*, \ldots, v_n^*\}$ is a basis for $\mathcal{V}_1^\perp$. This will imply that $\dim \mathcal{V}_1^\perp = n - m$, proving the proposition.

By (3.1.3) we have $\{v_{m+1}^*, \ldots, v_n^*\} \subset \mathcal{V}_1^\perp$ and, being part of a basis, these vectors are independent. We only need to prove that they span $\mathcal{V}_1^\perp$.

Let $w^* \in \mathcal{V}_1^\perp$, write $w^* = \sum_{j=1}^n a_j v_j^*$, and observe that $a_j = (v_j, w^*)$. If $w^* \in \mathcal{V}_1^\perp$, then $a_j = 0$ for $1 \leq j \leq m$, so that $w^* = \sum_{m+1}^n a_j v_j^*$. ◀

Corollary. *Let $\mathcal{V}_1 \subset \mathcal{V}$ be a subspace, and $v \in \mathcal{V} \setminus \mathcal{V}_1$. Then there exist $v^* \in \mathcal{V}_1^\perp$ such that $(v, v^*) \neq 0$.*

PROOF: As $\operatorname{span}[v, \mathcal{V}_1]$ properly contains $\mathcal{V}_1$, its dimension is bigger than that of $\mathcal{V}_1$. Its annihilator has lower dimension than $\mathcal{V}_1^\perp$, and is therefore *properly* contained in the latter. This is precisely our claim.
◀

Remark: The use of dimension to prove the corollary is somewhat heavy-handed and the proof is less transparent than the following observation. In the notation of the proof of the theorem, the assumption that $v \notin \mathcal{V}_1$ implies directly, by (3.1.4), that $(v, v_k^*) \neq 0$ for some $k > m$ (so that $v_k^* \in \mathcal{V}_1^\perp$).

Restating the corollary for the case that $\mathcal{V}_1$ is given as $\operatorname{span}[A]$ for some subset $A \subset \mathcal{V}$, we have

Proposition. *Let $A \subset \mathcal{V}$, $v \in \mathcal{V}$. Then $v \in \operatorname{span}[A]$ if and only if $(v, u^*) = 0$ for every $u^* \in A^\perp$.*

3.1.4 Let $\mathcal{V}$ be a finite-dimensional vector space and $\mathcal{V}_1 \subset \mathcal{V}$ a subspace. The restriction of a linear functional in $\mathcal{V}^*$ to $\mathcal{V}_1$ is a linear functional on $\mathcal{V}_1$.

The functionals whose restriction to $\mathcal{V}_1$ is zero are, by definition, the elements of $\mathcal{V}_1^\perp$. The restrictions of v^* and u^* to $\mathcal{V}_1$ are equal if and only if $v^* - u^* \in \mathcal{V}_1^\perp$. This, combined with exercise **ex3.1.2** below, gives a natural identification of $\mathcal{V}_1^*$ with the quotient space $\mathcal{V}^* / \mathcal{V}_1^\perp$.

EXERCISES FOR SECTION 3.1

ex3.1.1 Given a linearly independent $\{v_1, \ldots, v_k\} \subset \mathscr{V}$ and scalars $\{a_j\}_{j=1}^k$. Prove that there exists $v^* \in \mathscr{V}^*$ such that $(v_j, v^*) = a_j$ for $1 \le j \le k$.

ex3.1.2 If $\mathscr{V}_1$ is a subspace of a finite-dimensional space $\mathscr{V}$, then every linear functional on $\mathscr{V}_1$ is the restriction to $\mathscr{V}_1$ of a linear functional on $\mathscr{V}$.

ex3.1.3 Let $\mathscr{V}_1 \subset \mathscr{V}$ be a subspace. Prove that (identifying $(\mathscr{V}^*)^*$ with $\mathscr{V}$) $(\mathscr{V}_1^{\perp})^{\perp} = \mathscr{V}_1$.

ex3.1.4 Let $\mathscr{V}$ be a finite-dimensional vector space, and $\mathscr{V}_1 \subset \mathscr{V}$ a subspace. Let $\{u_k^*\}_{k=1}^r \subset \mathscr{V}^*$ be linearly independent mod $\mathscr{V}_1^{\perp}$ (that is: if $\sum c_k u_k^* \in \mathscr{V}_1^{\perp}$, then $c_k = 0$, for $k = 1, \ldots, r$). Let $\{v_j^*\}_{j=1}^s \subset \mathscr{V}_1^{\perp}$ be independent. Prove that $\{u_k^*\} \cup \{v_j^*\}$ is linearly independent in $\mathscr{V}^*$.

ex3.1.5 Show that every linear functional on $\mathbb{F}_c^n$ is given by some $(a_1, \ldots, a_n)$ in $\mathbb{F}_r^n$ as

$$\begin{bmatrix} x_1 \\ \vdots \\ x_n \end{bmatrix} \mapsto (a_1, \ldots, a_n) \begin{bmatrix} x_1 \\ \vdots \\ x_n \end{bmatrix} = \sum a_j x_j.$$

ex3.1.6 Let $\mathscr{V}$ and $\mathscr{W}$ be finite-dimensional vector spaces. Prove

a. For every $v \in \mathscr{V}$ and $w^* \in \mathscr{W}^*$ the map

$$\varphi_{v,w^*} : T \mapsto (Tv, w^*)$$

is a linear functional on $\mathscr{L}(\mathscr{V}, \mathscr{W})$.

b. The map $v \otimes w^* \mapsto \varphi_{v,w^*}$ extends to an isomorphism of $\mathscr{V} \otimes \mathscr{W}^*$ onto the dual space of $\mathscr{L}(\mathscr{V}, \mathscr{W})$.

ex3.1.7 Let $\mathscr{V}$ be a complex vector space, $\{v_j^*\}_{j=1}^s \subset \mathscr{V}^*$, and $w^* \in \mathscr{V}^*$ such that for all $v \in \mathscr{V}$,

$$|\langle v, w^* \rangle| \le \max_{j=1,\ldots,s} |\langle v, v_j^* \rangle|.$$

Prove that $w^* \in \text{span}[\{v_j^*\}_{j=1}^s]$.

ex3.1.8 *Linear functionals on* $\mathbb{R}_N[x]$:

a. Show that for every $x \in \mathbb{R}$ the map φ_x defined by $(P, \varphi_x) = P(x)$ is a linear functional on $\mathbb{R}_N[x]$.

b. If $\{x_1, \ldots, x_m\}$ are distinct and $m \le N + 1$, then φ_{x_j} are linearly independent.
   Hint: Write $P_l(x) = c_l \prod_{j \ne l}(x - x_j)$ with $c_l = \prod(x_l - x_j)^{-1}$.

c. Prove Lagrange's theorem: *Given m distinct numbers $\{x_1,\ldots,x_m\}$ and m arbitrary numbers $\{a_1,\ldots,a_m\}$, there exists a polynomial P of degree $m-1$ such that $P(x_j) = a_j$ for all j.*

d. For $x \in \mathbb{R}$ and $l \in \mathbb{N}$, $l \le N$, the map $\varphi_x^{(l)}$ defined by $(P, \varphi_x^{(l)}) = P^{(l)}(x)$ (the l'th derivative of P at x) is a nontrivial linear functional on $\mathbb{R}_N[x]$.

e. Let (x_j, l_j), $j = 1,\ldots,N+1$, $x_j \in \mathbb{R}$, $l_j \in \mathbb{N}$, be distinct pairs. Denote by $\#(m)$ the number of such pairs with $l_j > m$.

Prove that a necessary condition for the functionals $\varphi_{x_j}^{(l_j)}$ to be independent on $\mathbb{R}_N[x]$ is:

(3.1.6)   for every $m \le N$, $\#(m) \le N - m$.

Hint: If $l > k$, and x is arbitrary, then $\varphi_x^{(l)} \in (\mathbb{R}_k[x])^{\perp}$.

f. Check that φ_1, φ_{-1}, and $\varphi_0^{(1)}$ are linearly dependent in the dual of $\mathbb{R}_2[x]$, hence (3.1.6) is *not* sufficient. Are φ_1, φ_{-1}, and $\varphi_0^{(1)}$ linearly dependent in the dual of $\mathbb{R}_3[x]$?

ex3.1.9 Let $\mathscr{P} = \mathbb{R}_2[x,y,z]$, the space of quadratic polynomials in three variables with real-valued coefficients, that is,

(3.1.7) $\mathscr{P} = \{\sum a_{j,k,l} x^j y^k z^l : j \ge 0, k \ge 0, l \ge 0; \ j+k+l \le 2\}.$

Denote by ∂_u the directional derivative at $x = y = z = 0$ in the direction $u \in \mathbb{R}^3$, i.e., the map $P \mapsto \partial_u P(0)$. Denote by $\partial_{u,v}^2$ the corresponding mixed double derivative at the origin $x = y = z = 0$.

Prove that for all $u, v \in \mathbb{R}^3$, ∂_u and $\partial_{u,v}^2$ are linear functionals on $\mathscr{P}$.

Prove that if u, v, w are linearly independent in $\mathbb{R}^3$, the functionals
$$\partial_u, \ \partial_v, \ \partial_w, \ \partial_{u,u}^2, \ \partial_{u,v}^2, \ \partial_{u,w}^2, \ \partial_{v,v}^2, \ \partial_{v,w}^2, \ \partial_{w,w}^2$$
are linearly independent. Do they form a basis for $\mathscr{P}^*$?

3.2   The adjoint

3.2.1   Let $T \in \mathscr{L}(\mathscr{V}, \mathscr{W})$ and $w^* \in \mathscr{W}^* = \mathscr{L}(\mathscr{W}, \mathbb{F})$. The composition

(3.2.1) $$w^*T : v \mapsto (Tv, w^*)$$

is a linear map from $\mathscr{V}$ to the underlying field, i.e., a linear functional v^* on $\mathscr{V}$.

For T fixed, the map $T^*\colon w^* \mapsto w^*T$ from $\mathcal{W}^*$ into $\mathcal{V}^*$ is clearly linear. It is called the *adjoint* of T.

The basic relationship between T, T^*, and the bilinear forms (v,v^*) and (w,w^*) is: for all $v \in \mathcal{V}$ and $w^* \in \mathcal{W}^*$,

$$(3.2.2) \qquad (Tv, w^*) = (v, T^*w^*).$$

Notice that the bilinear form on the left-hand side is the bilinear form on (W, W^*), while on the right-hand side it is the one on (V, V^*).

If we identify the dual space $\mathcal{V}^{**}$ of $\mathcal{V}^*$ with $\mathcal{V}$ (see 3.1.2), and the dual $\mathcal{W}^{**}$ of $\mathcal{W}^*$ with $\mathcal{W}$, then (3.2.2) also identifies T as the adjoint T^{**} of T^*.

3.2.2 We have seen in 3.1.2 that if $\mathcal{V} = \mathbb{F}_{\mathbf{c}}^n$, $\mathcal{W} = \mathbb{F}_{\mathbf{c}}^m$, both with standard bases, then we can identify $\mathcal{V}^*$ with $\mathbb{F}_{\mathbf{r}}^n$ and $\mathcal{W}^*$ with $\mathbb{F}_{\mathbf{r}}^m$, and the standard basis of $\mathbb{F}_{\mathbf{r}}^n$ is the dual basis of the standard basis of $\mathbb{F}_{\mathbf{c}}^n$.

If $A = A_T = \begin{bmatrix} t_{11} & \cdots & t_{1n} \\ \vdots & \cdots & \vdots \\ t_{m1} & \cdots & t_{mn} \end{bmatrix}$ is the matrix of an operator $T \in \mathcal{L}(\mathbb{F}_{\mathbf{c}}^n, \mathbb{F}_{\mathbf{c}}^m)$

with respect to the standard bases, then the operator T is given as left-multiplication by A on $\mathbb{F}_{\mathbf{c}}^n$, and the bilinear form (Tv, w), for $w \in \mathbb{F}_{\mathbf{r}}^m$ and $v \in \mathbb{F}_{\mathbf{c}}^n$, is just the matrix product $w(Av)$. We have

$$(3.2.3) \qquad (Tv, w) = w(Av) = wAv = (wA)v,$$

and it follows that $T^*w = wA_T$. That means that the action of T^* on the row vectors in $\mathbb{F}_{\mathbf{r}}^m$ is obtained as right-multiplication by the *same matrix* $A = A_T$.

If we want[1] to have the matrix of T^* relative to the standard bases in $\mathbb{F}_{\mathbf{c}}^n$ and $\mathbb{F}_{\mathbf{c}}^m$, acting on columns by left-multiplication, all we need to do is transpose w and wA and obtain (see also example 2.4.8)

$$T^*w^{\mathbb{T}} = A^{\mathbb{T}}w^{\mathbb{T}}.$$

---

[1]This will be the case when there is a natural way to identify the vector space with its dual, for instance when we work with *inner-product spaces*. If the "identification" is through a sesquilinear form, as is the case when $\mathbb{F} = \mathbb{C}$, then the matrix for the adjoint is the complex conjugate of $A^{\mathbb{T}}$; see Chapter 6.

3.2.3 Proposition. *Let $T \in \mathscr{L}(\mathscr{V}, \mathscr{W})$, and let T^* be its adjoint. Then*

(3.2.4) $\operatorname{range}(T)^{\perp} = \ker(T^*)$ *and* $\operatorname{range}(T^*)^{\perp} = \ker(T)$.

PROOF: The condition $w^* \in \operatorname{range}(T)^{\perp}$ is equivalent to $(Tv, w^*) = 0$ for all $v \in \mathscr{V}$. Since $(Tv, w^*) = (v, T^*w^*)$, the condition is equivalent to $(v, T^*w^*) = 0$ for all $v \in \mathscr{V}$, which is equivalent to $T^*w^* = 0$. This proves the claim $\operatorname{range}(T)^{\perp} = \ker(T^*)$.

The second equality in (3.2.4) is the same statement as the first, applied to T^* and its adjoint $T^{**} = T$ instead of T and its adjoint T^*.
◀

Corollary. *Let $T \in \mathscr{L}(\mathscr{V}, \mathscr{W})$ and let T^* be its adjoint; then*

(3.2.5) $\rho(T^*) = \rho(T)$.

PROOF: Denote $n = \dim \mathscr{V} = \dim \mathscr{V}^*$ and $m = \dim \mathscr{W} = \dim \mathscr{W}^*$. Denote also $\rho(T) = \dim \operatorname{range}(T)$ and $\rho(T^*) = \dim \operatorname{range}(T^*)$.

By Theorem 3.1.3, $\dim \operatorname{range}(T)^{\perp} = m - \rho(T)$. By Theorem 2.5.1, $\dim \ker(T^*) = m - \rho(T^*)$. By (3.2.4) the two are equal.   ◀

EXERCISES FOR SECTION 3.2

ex3.2.1 If $\mathscr{V} = \mathscr{W} \oplus \mathscr{U}$ and S is the projection of $\mathscr{V}$ on $\mathscr{W}$ along $\mathscr{U}$ (see 2.1.1, example f), what is the adjoint S^*?

ex3.2.2 Show that Proposition 3.2.3 is equivalent to the statement that the column rank of a matrix is equal to its row rank.

ex3.2.3 A vector $v \in \mathscr{V}$ is an *eigenvector* for $T \in \mathscr{L}(\mathscr{V})$ if $Tv = \lambda v$ with $\lambda \in \mathbb{F}$; λ is the corresponding *eigenvalue*.

Let $v \in \mathscr{V}$ be an eigenvector of T with corresponding eigenvalue λ, and $w \in \mathscr{V}^*$ an eigenvector of the adjoint T^* with eigenvalue $\lambda^* \neq \lambda$. Prove that $(v, w^*) = 0$.

Chapter 4

Determinants

4.1 Permutations

As mentioned in example c of 1.1.1, a *permutation* of a set X is a bijective map of X onto itself, and the collection $S(X)$ of all the permutations of X forms a group under the operation of *composition* of maps: the product $\tau\sigma$ of the permutations $\sigma, \tau \in S(X)$ is defined by: for $j \in X$,

$$(\tau\sigma)(j) = \tau(\sigma(j)).$$

The identity element of $S(X)$ is the permutation e defined by $e(j) = j$ for all $j \in A$ (the *trivial permutation*).

We are interested mostly in permutations of the set $X = [1, \ldots, n]$. The group $S_n = S([1, \ldots, n])$ is called the *symmetric group on* $[1, \ldots, n]$.

4.1.1 For $\sigma \in S_n$, the *σ-orbit* of an element $a \in [1, \ldots, n]$ is the set $\{\sigma^k(a)\}$. If a is a *fixed point* of σ, i.e., $\sigma a = a$, its orbit is reduced to a single point; we refer to such orbits as *trivial*.

A *cycle* is a permutation with a unique nontrivial orbit. Cycles σ are often written as $(a_1, \ldots, a_m)$, where $\{a_j\}_{j=1}^m$ is the unique nontrivial orbit, enumerated in such a way that $a_{j+1} = \sigma(a_j)$ for $1 \leq j < m$, and $a_1 = \sigma(a_m)$.

Observe that σ is determined by the cyclic order of the entries, thus

$$(a_1, \ldots, a_m) = (a_m, a_1, \ldots, a_{m-1}).$$

Every orbit $O = \{\sigma^j a\}$ of a permutation σ determines a cycle τ_O by setting $\tau_O = \sigma$ on O, and $\tau_O = e$, the identity, on the complement $[1, \ldots, n] \setminus O$. We refer to τ_O as *the restriction of σ to O*.

Two cycles are said to be *disjoint* if their (nontrivial) orbits are disjoint.

Lemma. *Every $\sigma \in S_n$ is a product of pairwise disjoint cycles. The representation as such a product is unique.*

PROOF: The σ-orbits form a partition of $[1, \ldots, n]$, and σ is the product of its restrictions to the various orbits. For the uniqueness observe first that disjoint cycles commute; that means that the "uniqueness" must allow for arbitrary reordering of the factors. The claim is that the (unordered) set of factors is unique.

Now observe that if $\sigma = \prod \tau_j$, and the factors τ_j are pairwise disjoint cycles with orbits O_j, then the restriction of σ to O_j is equal to τ_j since all the factors τ_k, $k \neq j$, act like the identity on O_j. In other words, writing σ as the product of its restrictions to the various orbits is the only way to write it as a product of pairwise disjoint cycles. ◄

The σ-*period* of a point a is the number of points in its orbit; equivalently, it is the first positive integer l such that $\sigma^l(a) = a$. Notice that the *order* of a cycle $\sigma = (a_1, \ldots, a_m)$, that is, the smallest positive integer l such that σ^l is the identity, is equal to its length m, that is, the number of elements in its orbit.

4.1.2 Cycles of length 2 are called *transpositions*.

Lemma. *Every permutation $\sigma \in S_n$ is a product of transpositions.*

PROOF: Since every $\sigma \in S_n$ is a product of cycles, it suffices to show that every cycle is a product of transpositions. Observe that

$$(a_1, \ldots, a_l) = (a_l, a_1, a_2, \ldots, a_{l-1}) = (a_1, a_2)(a_2, a_3) \cdots (a_{l-1}, a_l).$$

In words: a_l trades places with a_{l-1}, then with a_{l-2}, etc., until it settles in place of a_1; every other a_j moves once, to the original place of a_{j+1}. Thus, every cycle of length l is a product of $l - 1$ transpositions. ◄

*$\textbf{4.1.3}$ **Conjugation in** S_n. If σ, $\tau \in S_n$, and $\tau(i) = j$ then $\tau\sigma^{-1}$ maps $\sigma(i)$ to j and $\sigma\tau\sigma^{-1}$ maps $\sigma(i)$ to $\sigma(j)$. This means that the cycles of $\sigma\tau\sigma^{-1}$ are obtained from the cycles of τ by replacing the entries in each (cycle of τ) by their σ-images. Conversely, if τ_1 and τ_2 have the same orbit structure, that is, have the same number of orbits of each length, and σ is a permutation that maps every orbit of τ_1 on a τ_2-orbit of the same length keeping the order of the corresponding cycles, we can verify, as above, that $\tau_2 = \sigma\tau_1\sigma^{-1}$. This proves the following proposition.

Proposition. *Two permutations are conjugate in S_n if and only if they have the same orbit structure. In particular, two cycles are conjugate in S_n if and only if they have the same length.*

$\textbf{4.1.4}$ **The sign of a permutation.** There are several equivalent ways to define *the sign* of a permutation $\sigma \in S_n$. The sign, denoted $\textbf{\textit{sgn}}\,[\sigma]$, is to take the values ± 1, assign the value -1 to each transposition, and be multiplicative: $\textbf{\textit{sgn}}\,[\sigma\tau] = \textbf{\textit{sgn}}\,[\sigma]\,\textbf{\textit{sgn}}\,[\tau]$; in other words, it is to be a homomorphism of S_n onto the multiplicative group $\{1, -1\}$.

All these requirements imply that if σ can be written as a product of k transpositions, then $\textbf{\textit{sgn}}\,[\sigma] = (-1)^k$. But in order to use this as the *definition* of $\textbf{\textit{sgn}}$, one needs to prove that the numbers of factors in all the representations of any $\sigma \in S_n$ as products of transpositions have the same parity.

We introduce $\textbf{\textit{sgn}}$ in a different way:

DEFINITION: We say that a set J of pairs $\{(k, l)\}$ of distinct integers from $[1, \dots, n]$ is *appropriate for S_n* if it contains exactly one of (j, i), (i, j) for every pair i, j, $1 \le i < j \le n$.

A simple example is $J = \{(i, j) : 1 \le i < j \le n\}$. A more general example of an appropriate set is: for $\tau \in S_n$,

(4.1.1) $\qquad J_\tau = \{(\tau(i), \tau(j)) : 1 \le i < j \le n\}$.

If J is appropriate for S_n, and $\sigma \in S_n$, then[1]

(4.1.2) $\quad \displaystyle\prod_{i<j} \mathrm{sgn}\,(\sigma(j) - \sigma(i)) = \prod_{(i,j) \in J} \mathrm{sgn}\,(\sigma(j) - \sigma(i))\,\mathrm{sgn}\,(j - i)$

---

[1] The sign of integers has the usual meaning.

since reversing the order of a pair (i, j) changes both $\text{sgn}\,(\sigma(j) - \sigma(i))$
and $\text{sgn}\,(j - i)$, and does not affect their product. Notice that with
$J = J_\tau$ of (4.1.1) this takes the form

$$\prod_{i<j} \text{sgn}\,(\sigma(j) - \sigma(i)) = \prod_{i<j} \text{sgn}\,(\sigma\tau(j) - \sigma\tau(i))\,\text{sgn}\,(\tau(j) - \tau(i)).$$

We define the sign of a permutation σ by

(4.1.3) $$\boldsymbol{sgn}\,[\sigma] = \prod_{i<j} \text{sgn}\,(\sigma(j) - \sigma(i)).$$

Proposition. *The map* $\boldsymbol{sgn}$: $\sigma \mapsto \boldsymbol{sgn}\,[\sigma]$ *is a homomorphism of* S_n
onto the multiplicative group $\{1, -1\}$. *The sign of any transposition is*
-1.

PROOF: The multiplicativity is seen as follows:

$$\boldsymbol{sgn}\,[\sigma\tau] = \prod_{i<j} \text{sgn}\,(\sigma\tau(j) - \sigma\tau(i))$$

$$= \prod_{i<j} \text{sgn}\,(\sigma\tau(j) - \sigma\tau(i))\,\text{sgn}\,(\tau(j) - \tau(i)) \prod_{i<j} \text{sgn}\,(\tau(j) - \tau(i))$$

$$= \boldsymbol{sgn}\,[\sigma]\,\boldsymbol{sgn}\,[\tau].$$

Since the sign of the identity permutation is $+1$, the multiplicativity implies that conjugate permutations have the same sign. In particular all transpositions have the same sign. The computation for $(1, 2)$ is particularly simple:
$\text{sgn}\,(j - 1) = \text{sgn}\,(j - 2) = 1$ for all $j > 2$, while $\text{sgn}\,(1 - 2) = -1$,
and it follows that the sign of every transposition is -1.   ◄

EXERCISES FOR SECTION 4.1

ex4.1.1 How many cycles of length n are there in S_n? How many cycles of
length m are there in S_n?

ex4.1.2 Let $\sigma_j \in S_n$, $j = 1, 2$, be cycles with different orbits. Prove that the
two commute if and only if their (nontrivial) orbits are disjoint.

ex4.1.3 Let σ, τ be cycles of length l with the same nontrivial orbit. Prove that the two commute if and only if each is a power of the other: $\sigma = \tau^k$ and $\tau = \sigma^m$. Can either k or m have a common factor with l?

Hint: $km - 1$ is divisible by l.

ex4.1.4 Let σ be a cycle of length k; prove that $\textit{\textbf{sgn}}\,[\sigma] = (-1)^{k-1}$.

ex4.1.5 Let $\sigma \in S_n$ and assume that it has s orbits (including the trivial orbits, i.e., fixed points). Prove that $\textit{\textbf{sgn}}\,[\sigma] = (-1)^{n-s}$

4.2 Multilinear maps

4.2.1 Let $\mathscr{V}_j$, $j = 1, \ldots, k$, and $\mathscr{W}$ be vector spaces over a field $\mathbb{F}$.

A map

$$(4.2.1) \qquad \psi : \mathscr{V}_1 \times \mathscr{V}_2 \times \cdots \times \mathscr{V}_k \to \mathscr{W}$$

is *multilinear*, or k-linear (*bilinear*—if $k = 2$) if $\psi(v_1, \ldots, v_k)$ is linear in each entry v_j when the other entries are held fixed.

When all the $\mathscr{V}_j$'s are equal to some fixed $\mathscr{V}$, we say that ψ is k-linear on $\mathscr{V}$. If $\mathscr{W}$ is the underlying field $\mathbb{F}$, we refer to ψ as a *k-linear form* or just *k-form*.

EXAMPLES:

a. Multiplication in an algebra, for example, $(S, T) \mapsto ST$ in $\mathscr{L}(\mathscr{V})$ or $(A, B) \mapsto AB$ in $\mathscr{M}(n)$.

b. Let $\mathscr{V}_1 = \mathbb{F}[x]$ and $\mathscr{V}_2 = \mathbb{F}[y]$; then the map $(p(x), q(y)) \mapsto p(x)q(y)$ is a bilinear map from $\mathbb{F}[x] \times \mathbb{F}[y]$ onto the space $\mathbb{F}[x, y]$ of polynomials in two variables.

c. $\psi(v, v^*) = (v, v^*)$, the value of a linear functional $v^* \in \mathscr{V}^*$ on a vector $v \in \mathscr{V}$, is a bilinear form on $\mathscr{V} \times \mathscr{V}^*$.

d. Given k linear functionals $v_j^* \in \mathscr{V}^*$, their product

$$(4.2.2) \qquad \psi_{v_1^*, \ldots, v_k^*}(v_1, \ldots, v_k) = \prod (v_j, v_j^*)$$

is a k-form on $\mathscr{V}$.

4.2.2   If Ψ and Φ are k-linear maps of $\mathscr{V}_1 \times \mathscr{V}_2 \times \cdots \times \mathscr{V}_k$ into $\mathscr{W}$ and $a, b \in \mathbb{F}$, then $a\Psi + b\Phi$ is k-linear. It follows that the k-linear maps of $\mathscr{V}_1 \times \mathscr{V}_2 \times \cdots \times \mathscr{V}_k$ into $\mathscr{W}$ form a vector space, which we denote by $\mathscr{ML}(\{\mathscr{V}_j\}_{j=1}^k, \mathscr{W})$. When all the $\mathscr{V}_j$ are the same space $\mathscr{V}$, the notation is: $\mathscr{ML}(\mathscr{V}^{\oplus k}, \mathscr{W})$, and when $\mathscr{W} = \mathbb{F}$, the reference to $\mathscr{W}$ is omitted. Thus, $\mathscr{ML}(\mathscr{V}^{\oplus k})$ is the space of all k-forms on $\mathscr{V}$.

4.2.3   Example d of 4.2.1 is very useful in that the linear span of the forms it defines is the space of all k-forms on $\mathscr{V}$.

Theorem. *Assume that $\mathscr{V}$ is finite-dimensional. Let $\{u_1, \ldots, u_n\}$ be a basis for $\mathscr{V}$, and $\mathbf{u}^* = \{u_1^*, \ldots, u_n^*\}$ the dual basis. Let the forms $\psi_{v_1^*, \ldots, v_k^*}$ be defined by (4.2.2). Then the set*

(4.2.3) $\{\psi_{v_1^*, \ldots, v_k^*} : v_m^* \in \mathbf{u}^*, \ m = 1, \ldots, k\}$

spans $\mathscr{ML}(\mathscr{V}^{\oplus k})$.

PROOF: We use induction on k. If $k = 1$ there is nothing to prove. Assume the result valid for $k - 1$. Let φ be a k-form. Write

(4.2.4) $\varphi_j(v_2, \ldots, v_k) = \varphi(u_j, v_2, \ldots, v_k)$

and observe that for every v_1 (as $v_1 = \sum (v_1, u_j^*) u_j$),

(4.2.5) $\varphi(v_1, v_2, \ldots, v_k) = \sum (v_1, u_j^*) \varphi_j(v_2, \ldots, v_k).$

The induction hypothesis applies to the $k - 1$-forms φ_j, and expressing φ_j in (4.2.5) as a linear combination of $\psi_{v_1^*, \ldots, v_{k-1}^*}$ completes the proof.
◀

*4.2.4   The definition in 1.2.6 of the tensor product $\mathscr{V}_1 \otimes \mathscr{V}_2$ guarantees that the map

(4.2.6) $\Psi(v, u) = v \otimes u$

of $\mathscr{V}_1 \times \mathscr{V}_2$ into $\mathscr{V}_1 \otimes \mathscr{V}_2$ is bilinear. This map is special in that every bilinear map from $(\mathscr{V}_1, \mathscr{V}_2)$ factors through it.

Theorem. *Let φ be a bilinear map from $(\mathscr{V}_1, \mathscr{V}_2)$ into $\mathscr{W}$. Then there is a linear map $\Phi \colon \mathscr{V}_1 \otimes \mathscr{V}_2 \longrightarrow \mathscr{W}$ such that $\varphi = \Phi \Psi$.*

PROOF: The operator Φ is defined by: $\Phi(v \otimes u) = \varphi(v, u)$. This is unambiguous since for $v_j \in \mathscr{V}_1$ and $u_j \in \mathscr{V}_2$,

$$\sum v_j \otimes u_j = 0 \implies \sum \varphi(v_j, u_j) = 0.$$

One needs only to check that, so defined, the map Φ is linear, and that $\varphi = \Phi \Psi$. We leave this to the reader. ◀

∗**4.2.5** Let $\mathscr{V}$ and $\mathscr{W}$ be finite-dimensional vector spaces.

If $T \in \mathscr{L}(\mathscr{V}, \mathscr{W})$ is of rank 1, and $w \neq 0$ is in the range of T, we have, for all $v \in \mathscr{V}$, $Tv = a(v)w$ and the coefficient $a(v)$ is a linear form. In other words: there exists $v^* \in \mathscr{V}^*$ such that $Tv = (v, v^*)w$.

On the other hand, for any choice of $v^* \in \mathscr{V}^*$ and $w \in \mathscr{W}$ the map $\phi_{v^* \otimes w}$ defined by $\phi_{v^* \otimes w} v = (v, v^*)w$ is clearly a linear operator of rank one from $\mathscr{V}$ to $\mathscr{W}$.

Theorem. *The map $\Phi \colon v^* \otimes w \mapsto \phi_{v^* \otimes w} \in \mathscr{L}(\mathscr{V}, \mathscr{W})$ extends by linearity to an isomorphism of $\mathscr{V}^* \otimes \mathscr{W}$ onto $\mathscr{L}(\mathscr{V}, \mathscr{W})$.*

PROOF: As in ∗4.2.4 we see that all the representations of zero in the tensor product are mapped to 0, so that Φ can be extended unambiguously to a linear map defined on all of $\mathscr{V}^* \otimes \mathscr{W}$. Since the two spaces have the same dimension, it is sufficient to show that Φ is surjective.

So let $T \in \mathscr{L}(\mathscr{V}, \mathscr{W})$, $\mathbf{v} = \{v_j\}$ a basis for $\mathscr{V}$, and $\mathbf{v}^* = \{v_j^*\}$ the dual basis. Then, for $v \in \mathscr{V}$,

$$(4.2.7) \qquad Tv = T\left(\sum (v, v_j^*) v_j\right) = \sum (v, v_j^*) Tv_j = \left(\sum \phi_{v_j^* \otimes Tv_j}\right) v,$$

so that $T = \sum \phi_{v_j^* \otimes Tv_j}$. This shows that Φ is surjective. ◀

When there is no room for confusion, we simplify the notation and write the operator as $v^* \otimes w$ instead of $\phi_{v^* \otimes w}$.

4.2.6 Symmetric and alternating k-forms. Let $\mathscr{V}$ be a vector space of dimension n, and let $\mathscr{V}^*$ be its dual.

A bilinear form φ on $\mathscr{V}$ is *symmetric* if $\varphi(v_1, v_2) = \varphi(v_2, v_1)$ for all $v_1, v_2 \in \mathscr{V}$; it is *alternating* if $\varphi(v, v) = 0$ for all $v \in \mathscr{V}$.

Example: given a bilinear form ψ, define

(4.2.8)
$$\psi_{sym}(v_1, v_2) = \psi(v_1, v_2) + \psi(v_2, v_1),$$
$$\psi_{alt}(v_1, v_2) = \psi(v_1, v_2) - \psi(v_2, v_1);$$

then the form ψ_{sym} is symmetric, while ψ_{alt} is alternating.

More generally, a k-form φ is *symmetric* if

(4.2.9)
$$\varphi(v_{\sigma(1)}, \ldots, v_{\sigma(k)}) = \varphi(v_1, \ldots, v_k).$$

for every k vectors $v_1, \ldots, v_k \in \mathscr{V}$ and every permutation $\sigma \in S_k$.

A k-form φ is *alternating*, if $\varphi(v_1, \ldots, v_k) = 0$ whenever two of the entries are equal, i.e.,

(4.2.10)   If $v_j = v_l$ for $j \neq l$, then $\varphi(v_1, \ldots, v_k) = 0$.

Condition (4.2.10) is equivalent to the seemingly stronger condition:

(4.2.11)
$$\varphi(v_{\sigma(1)}, \ldots, v_{\sigma(k)}) = \textbf{\textit{sgn}}\,[\sigma]\varphi(v_1, \ldots, v_k),$$

(for every k vectors $v_1, \ldots, v_k \in \mathscr{V}$ and every permutation $\sigma \in S_k$).

To show the equivalence we observe first that, since every permutation is a product of transpositions, and since $\textbf{\textit{sgn}}\,[\sigma]$ is multiplicative on S_k (see 4.1) it is enough to verify (4.2.11) for all transpositions (i, j). Using (4.2.10) and multilinearity, we have

(4.2.12)
$$\varphi(\ldots, v_i, \ldots, v_j, \ldots) = \varphi(\ldots, v_i, \ldots, v_j + v_i, \ldots)$$
$$= \varphi(\ldots, -v_j, \ldots, v_j + v_i, \ldots) = \varphi(\ldots, -v_j, \ldots, v_i, \ldots)$$
$$= -\varphi(\ldots, v_j, \ldots, v_i, \ldots),$$

proving (4.2.11) for transpositions, hence for all permutations.

The sets $\mathscr{ML}_{sym}(\mathscr{V}^{\oplus k})$ of all symmetric k-forms and $\mathscr{ML}_{alt}(\mathscr{V}^{\oplus k})$ of all alternating k-forms are linear subspaces of $\mathscr{ML}(\mathscr{V}^{\oplus k})$.

EXERCISES FOR SECTION 4.2

ex4.2.1 Refine Theorem 4.2.3 and show that if $\{u_1^*, \ldots, u_n^*\}$ is a basis for $\mathscr{V}^*$, then $\{\psi_{v_1^*, \ldots, v_k^*} : v_m^* \in \{u_1^*, \ldots, u_n^*\}, m = 1, \ldots, k\}$ is a basis for $\mathscr{ML}(\mathscr{V}^{\oplus k})$.
Hint: $\prod_{j=1}^{k} (u_{j_m}, u_{l_m}^*) = 1$ if $j_m = l_m$ for all m and is zero otherwise.

ex4.2.2 Prove that the sets $\mathscr{ML}_{sym}(\mathscr{V}^{\oplus 2})$ of symmetric bilinear forms on $\mathscr{V}$ and $\mathscr{ML}_{alt}(\mathscr{V}^{\oplus 2})$ of alternating ones are linear subspaces of $\mathscr{ML}(\mathscr{V}^{\oplus 2})$, and

$$(4.2.13) \qquad \mathscr{ML}(\mathscr{V}^{\oplus 2}) = \mathscr{M}_{sym}\mathscr{L}(\mathscr{V}^{\oplus 2}) \oplus \mathscr{ML}_{alt}(\mathscr{V}^{\oplus 2}).$$

ex4.2.3 If φ is a k-form on $\mathscr{V}$ and φ_{sym} is defined by

$$(4.2.14) \qquad \varphi_{sym}(v_1, \ldots, v_k) = \frac{1}{k!} \sum_{\sigma \in S_k} \varphi(v_{\sigma(1)}, \ldots v_{\sigma(k)}),$$

then φ_{sym} is symmetric, and the map $\varphi \mapsto \varphi_{sym}$ is a projection of $\mathscr{ML}(\mathscr{V}^{\oplus k})$ onto $\mathscr{ML}_{sym}(\mathscr{V}^{\oplus k})$. Similarly, the form φ_{alt} defined by

$$(4.2.15) \qquad \varphi_{alt}(v_1, \ldots, v_k) = \frac{1}{k!} \sum_{\sigma \in S_k} sgn\,[\sigma] \varphi(v_{\sigma(1)}, \ldots, v_{\sigma(k)}),$$

is alternating, and the map $\varphi \mapsto \varphi_{alt}$ is a projection of $\mathscr{ML}(\mathscr{V}^{\oplus k})$ onto the subspace $\mathscr{ML}_{alt}(\mathscr{V}^{\oplus k})$.

ex4.2.4 Let $\{e_i\}_{1 \leq i \leq n}$ be a basis of $\mathscr{V}^*$. For $i < j$, let $e_i \wedge e_j$ be the alternating bilinear forms

$$e_i \wedge e_j(v_1, v_2) = (v_1, e_i)(v_2, e_j) - (v_1, e_j)(v_2, e_i)$$

on $\mathscr{V} \times \mathscr{V}$. Prove that $\{e_i \wedge e_j\}_{i<j}$ is a basis for $\mathscr{ML}_{alt}(\mathscr{V}^{\oplus 2})$.

What are the dimensions of the spaces

$$\mathscr{ML}(\mathscr{V}^{\oplus 2}), \quad \mathscr{ML}_{sym}(\mathscr{V}^{\oplus 2}), \quad \text{and} \quad \mathscr{ML}_{alt}(\mathscr{V}^{\oplus 2}).$$

ex4.2.5 With $\{e_i\}_{1 \leq i \leq n}$ as above and for $i < j < k$, let $e_i \wedge e_j \wedge e_k$ be the trilinear forms

$$(4.2.16) \quad e_i \wedge e_j \wedge e_k(v_1, v_2, v_3) = \sum_{\sigma \in S_3} sgn\,[\sigma](v_{\sigma(1)}, e_i)(v_{\sigma(2)}, e_j)(v_{\sigma(3)}, e_k).$$

Prove that each $e_i \wedge e_j \wedge e_k$ is alternating and show that the set

$$\{e_i \wedge e_j \wedge e_k : i < j < k\}$$

is a basis for $\mathcal{ML}_{alt}(\mathcal{V}^{\oplus 3})$.

ex4.2.6 Assume that $\varphi(v, u)$ is bilinear on $\mathcal{V}_1 \times \mathcal{V}_2$.

Prove that the map $T : u \mapsto \varphi(\cdot, u)$ is a linear map from $\mathcal{V}_2$ into (the dual space) $\mathcal{V}_1^*$. Similarly, $S : v \mapsto \varphi(v, \cdot)$ is linear from $\mathcal{V}_1$ to $\mathcal{V}_2^*$.

ex4.2.7 Let $\mathcal{V}_1$ and $\mathcal{V}_2$ be finite-dimensional, with bases $\{v_1, \dots, v_m\}$ and $\{u_1, \dots, u_n\}$ respectively. Show that every bilinear form φ on $(\mathcal{V}_1, \mathcal{V}_2)$ is given by an $m \times n$ matrix (a_{jk}) such that if $v = \sum_1^m x_j v_j$ and $u = \sum_1^n y_k u_k$ then

$$(4.2.17) \qquad \varphi(v, u) = \sum a_{jk} x_j y_k = [x_1, \dots, x_m] \begin{bmatrix} a_{11} & \cdots & a_{1n} \\ \vdots & \cdots & \vdots \\ a_{m1} & \cdots & a_{mn} \end{bmatrix} \begin{bmatrix} y_1 \\ \vdots \\ y_n \end{bmatrix}.$$

ex4.2.8 What is the relation between the matrix in **ex4.2.7** and the maps S and T defined in **ex4.2.6**?

ex4.2.9 Assume that $\mathcal{V}_1$ and $\mathcal{V}_2$ are finite-dimensional, with bases $\{v_1, \dots, v_m\}$ and $\{u_1, \dots, u_n\}$ respectively. Let $T \in \mathcal{L}(\mathcal{V}_1, \mathcal{V}_2)$ and let

$$A_T = \begin{bmatrix} a_{11} & \cdots & a_{1m} \\ \vdots & \cdots & \vdots \\ a_{n1} & \cdots & a_{nm} \end{bmatrix}$$

be its matrix relative to the given bases. Let $\{v_1^*, \dots, v_m^*\}$ be the dual basis of $\{v_1, \dots, v_m\}$. Prove that

$$(4.2.18) \qquad T = \sum a_{ij}(v_i^* \otimes u_j).$$

4.3   Alternating n-forms

4.3.1   If φ is an alternating n-form, and if one of the entry vectors in $\varphi(v_1, \dots, v_n)$ is a linear combination of the others, we use the linearity of φ in that entry and write $\varphi(v_1, \dots, v_n)$ as a linear combination of φ evaluated on several n-tuples each of which has a repeated entry. Thus, if $\{v_1, \dots, v_n\}$ is linearly dependent, $\varphi(v_1, \dots, v_n) = 0$. It follows that if $\dim \mathcal{V} < n$, there are no nontrivial alternating n-forms on $\mathcal{V}$.

Theorem. *Assume that* $\dim \mathscr{V} = n$. *The space of alternating n-forms on $\mathscr{V}$ is one-dimensional: up to multiplication by a scalar there exists a unique nontrivial alternating n-form D on $\mathscr{V}$.*

Moreover, $D(v_1,\ldots,v_n) \neq 0$ if and only if $\{v_1,\ldots,v_n\}$ is a basis.

PROOF: Assume that $\dim \mathscr{V} = n$, let $\mathbf{u} = \{u_1,\ldots,u_n\}$ be a basis for $\mathscr{V}$ and let $\{u_1^*,\ldots,u_n^*\}$ be the dual basis. The alternating n-form

$$(4.3.1) \qquad D_{\mathbf{u}}(v_1,\ldots,v_n) = \sum_{\sigma \in S_n} sgn\,[\sigma] \prod (v_{\sigma(j)}, u_j^*)$$

is nontrivial since, as $\prod (u_{\sigma(j)}, u_j^*) = 1$ when σ is the identity and $\prod (u_{\sigma(j)}, u_j^*) = 0$ otherwise, $D(u_1,\ldots,u_n) = 1$.

It remains to show that if φ is an alternating n-form, then it is a scalar multiple of $D_{\mathbf{u}}$. Specifically, we must show that

$$(4.3.2) \qquad \varphi(v_1,\ldots,v_n) = \varphi(u_1,\ldots,u_n)D_{\mathbf{u}}(v_1,\ldots,v_n).$$

Let φ be an alternating n-form, then $\varphi(u_{j_1},\ldots,u_{j_n}) = 0$ if there is a repeated index, that is, unless $\{j_1,\ldots,j_n\}$ is a permutation σ of $\{1,\ldots,n\}$, and then $\varphi(u_{\sigma(1)},\ldots,u_{\sigma(n)}) = sgn\,[\sigma]\varphi(u_1,\ldots,u_n)$.

If $\{v_1,\ldots,v_n\}$ is an arbitrary n-tuple, we express each v_i in terms of the basis $\{u_1,\ldots,u_n\}$:

$$(4.3.3) \qquad v_j = \sum_{i=1}^{n} a_{i,j}u_i, \quad j = 1,\ldots,n,$$

where $a_{i,j} = (v_j, u_i^*)$ and the alternating multilinearity implies

$$(4.3.4) \qquad \begin{aligned} \varphi(v_1,\ldots,v_n) &= \sum a_{1,j_1} \cdots a_{n,j_n} \varphi(u_{j_1},\ldots,u_{j_n}) \\ &= \left(\sum_{\sigma \in S_n} sgn\,[\sigma] a_{1,\sigma(1)} \cdots a_{n,\sigma(n)} \right) \varphi(u_1,\ldots,u_n) \\ &= D_{\mathbf{u}}(v_1,\ldots,v_n)\varphi(u_1,\ldots,u_n). \end{aligned}$$

This shows that the value of $\varphi(u_1,\ldots,u_n)$ for a basis $\{u_1,\ldots,u_n\}$ determines $\varphi(v_1,\ldots,v_n)$ for all n-tuples. It also shows that all alternating n-forms are proportional and that, unless φ is trivial, $\varphi(v_1,\ldots,v_n) \neq 0$ for every independent set (i.e., basis) $\{v_1,\ldots,v_n\}$. ◀

Notice that even though the coefficients $a_{i,j}$ in (4.3.3) depend on the choice of the basis $\{u_1, \ldots, u_n\}$ used, the value of the expression

$$D_{\mathbf{u}}(v_1, \ldots, v_n) = \sum_{\sigma \in S_n} sgn\,[\sigma] a_{1,\sigma(1)} \cdots a_{n,\sigma(n)}$$

depends only on the value of $\varphi(u_1, \ldots, u_n)$, where φ is an arbitrary alternating n-form.

4.4   Determinant of an operator

4.4.1   Let D be a nontrivial alternating n-form on an n-dimensional vector space $\mathscr{V}$, and let $\{v_1, \ldots, v_n\}$ be a basis for $\mathscr{V}$.

DEFINITION: The *determinant* $\det T$ of an operator $T \in \mathscr{L}(\mathscr{V})$ is

$$(4.4.1) \qquad \det T = \frac{D(Tv_1, \ldots, Tv_n)}{D(v_1, \ldots, v_n)}.$$

The definition does not depend on the choice of D; another choice is just a nonzero constant multiple which appears in both the numerator and the denominator, and hence does not affect the quotient.

The definition is also independent of the choice of the basis. If T is singular, then $\{Tv_1, \ldots, Tv_n\}$ is linearly dependent (for any basis) and the determinant is 0. If T is nonsingular and $\{w_1, \ldots, w_n\}$ is another basis, express w_j in terms of $\{v_1, \ldots, v_n\}$:

$$(4.4.2) \qquad w_j = \sum c_{i,j} v_j, \quad j = 1, \ldots, n,$$

and observe that

$$(4.4.3) \qquad Tw_j = \sum c_{i,j} Tv_j, \quad j = 1, \ldots, n,$$

and the calculation in equation (4.3.4) shows that

$$(4.4.4) \qquad \frac{D(Tw_1, \ldots, Tw_n)}{D(Tv_1, \ldots, Tv_n)} = \frac{D(w_1, \ldots, w_n)}{D(v_1, \ldots, v_n)},$$

which implies that using the basis $\{w_1, \ldots, w_n\}$ would give the same value as (4.4.1).

If $A_{T,\mathbf{v}} = (a_{i,j})$ is the matrix of T in terms of a basis $\mathbf{v} = \{v_1, \ldots, v_n\}$, then

$$(4.4.5) \qquad Tv_j = \sum_{i=1}^{n} a_{i,j} v_i, \quad j = 1, \ldots, n,$$

and

$$(4.4.6) \qquad \det T = \sum_{\sigma \in S_n} \textbf{\textit{sgn}}\,[\sigma] a_{1,\sigma(1)} \cdots a_{n,\sigma(n)}.$$

Notice that while the matrix $A_{T,\mathbf{v}}$ depends on the choice of $\mathbf{v}$, the determinant does not.

4.4.2 Proposition. *Let $T \in \mathscr{L}(\mathscr{V})$. Then $\det T = 0$ if and only if T is singular.*

PROOF: This was mentioned in the proof that $\det T$ is independent of the basis used in its definition. T is singular if and only if it maps a basis onto a linearly dependent set, and $D(Tv_1, \ldots, Tv_n) = 0$ if and only if $\{Tv_1, \ldots, Tv_n\}$ is linearly dependent. ◄

4.4.3 Proposition. *If $T, S \in \mathscr{L}(\mathscr{V})$ then*

$$(4.4.7) \qquad \det TS = \det T \det S.$$

PROOF: If either S or T is singular, then both sides of (4.4.7) are zero. Otherwise $\det S \neq 0$, $\{Sv_j\}$ is a basis, and by (4.4.1),

$$\det TS = \frac{D(TSv_1, \ldots, TSv_n)}{D(Sv_1, \ldots, Sv_n)} \cdot \frac{D(Sv_1, \ldots, Sv_n)}{D(v_1, \ldots, v_n)} = \det T \det S.$$
◄

∗**4.4.4 Orientation.** When $\mathscr{V}$ is a real vector space, a nontrivial alternating n-form D determines an equivalence relation among bases. The bases $\{v_j\}$ and $\{u_j\}$ are declared equivalent if $D(v_1, \ldots, v_n)$ and $D(u_1, \ldots, u_n)$ have the same sign. Using $-D$ instead of D reverses the signs of all the readings, but maintains the equivalence. An *orientation* on $\mathscr{V}$ is a choice which of the two equivalence classes to call *positive*.

4.4.5   Invariant subspaces.   Let $T \in \mathscr{L}(\mathscr{V})$.

DEFINITION: A subspace $\mathscr{W} \subset \mathscr{V}$ is T-*invariant* if $w \in \mathscr{W}$ implies
$Tw \in \mathscr{W}$. If $\mathscr{W}$ is T-invariant , the *restriction* $T_{\mathscr{W}}$, defined by: $w \mapsto Tw$
for $w \in \mathscr{W}$, is clearly a linear operator on $\mathscr{W}$.

T also induces an operator $T_{\mathscr{V}/\mathscr{W}}$ on the quotient space $\mathscr{V}/\mathscr{W}$. The
operator $T_{\mathscr{V}/\mathscr{W}}$ is defined by

(4.4.8) $$T_{\mathscr{V}/\mathscr{W}}(v + \mathscr{W}) = Tv + \mathscr{W}.$$

The coset $v + \mathscr{W}$ of a vector v is mapped onto the coset of Tv. The
definition is justified by showing that it is independent of the choice
of the representative: if v_1 and v_2 represent the same coset, that is,
$v_1 - v_2 \in \mathscr{W}$, then, as $\mathscr{W}$ is T-invariant, $Tv_1 - Tv_2 = T(v_1 - v_2) \in \mathscr{W}$
so that Tv_1 and Tv_2 represent (belong to) the same coset.

The operator $T_{\mathscr{V}/\mathscr{W}}$ is linear: write $\tilde{v}$ for the coset $v + \mathscr{W}$; then
$T_{\mathscr{V}/\mathscr{W}}(a_1\tilde{v}_1 + a_2\tilde{v}_2)$ is the coset containing $a_1 Tv_1 + a_2 Tv_2$, which is the
sum for $j = 1,2$ of the cosets $a_j Tv_j + \mathscr{W}$, i.e., $a_1 T_{\mathscr{V}/\mathscr{W}}\tilde{v}_1 + a_2 T_{\mathscr{V}/\mathscr{W}}\tilde{v}_2$.

Proposition. *If $\mathscr{W} \subset \mathscr{V}$ is T-invariant, then*

(4.4.9) $$\det T = \det T_{\mathscr{W}} \det T_{\mathscr{V}/\mathscr{W}}.$$

PROOF: Let $\{w_j\}_1^n$ be a basis for $\mathscr{V}$, such that $\{w_j\}_1^k$ is a basis for $\mathscr{W}$.
If $T_{\mathscr{W}}$ is singular, then T is singular and both sides of (4.4.9) are zero.

If $T_{\mathscr{W}}$ is nonsingular, then $\mathbf{w} = \{Tw_1, \dots, Tw_k\}$ is a basis for $\mathscr{W}$,
and $\{Tw_1, \dots, Tw_k; w_{k+1}, \dots, w_n\}$ is a basis for $\mathscr{V}$.

Let D be a nontrivial alternating n-form on $\mathscr{V}$. Then

$$\Phi(u_1, \dots, u_k) = D(u_1, \dots, u_k; w_{k+1}, \dots, w_n)$$

is a nontrivial alternating k-form on $\mathscr{W}$.

The value of $D(Tw_1, \dots, Tw_k; u_{k+1}, \dots, u_n)$ is left unchanged if the
variables $u_{k+1}, \dots, u_n$ are replaced by ones that are congruent to them
mod $\mathscr{W}$. In other words, denoting by $\tilde{u}$ the equivalence class (mod $\mathscr{W}$)
of vectors u, the form $\Psi(\tilde{u}_{k+1}, \dots, \tilde{u}_n) = D(Tw_1, \dots, Tw_k; u_{k+1}, \dots, u_n)$

is therefore a well-defined nontrivial alternating $n-k$-form on $\mathscr{V}/\mathscr{W}$:

$$
\begin{aligned}
\det T &= \frac{D(Tw_1,\ldots,Tw_n)}{D(w_1,\ldots,w_n)} \\
&= \frac{D(Tw_1,\ldots,Tw_k;w_{k+1},\ldots,w_n)}{D(w_1,\ldots,w_n)} \cdot \frac{D(Tw_1,\ldots,Tw_n)}{D(Tw_1,\ldots,Tw_k;w_{k+1},\ldots,w_n)} \\
&= \frac{\Phi(Tw_1,\ldots,Tw_k)}{\Phi(w_1,\ldots,w_k)} \cdot \frac{\Psi(\widetilde{Tw}_{k+1},\ldots,\widetilde{Tw}_n)}{\Psi(\tilde{w}_{k+1},\ldots,\tilde{w}_n)} = \det T_{\mathscr{W}} \det T_{\mathscr{V}/\mathscr{W}}.
\end{aligned}
$$

◀

A special case of the proposition is seen in the following situation. Assume that $T \in \mathscr{L}(\mathscr{V})$, that $\mathscr{V} = \mathscr{V}_1 \oplus \mathscr{V}_2$, and that both components are T-invariant. One can then identify $\mathscr{V}_2$ with $\mathscr{V}/\mathscr{V}_1$ and $T_{\mathscr{V}_2}$ with $T_{\mathscr{V}/\mathscr{V}_1}$, and the proposition reads

(4.4.10) $$\det T = \det T_{\mathscr{V}_1} \det T_{\mathscr{V}_2}.$$

By induction on the number of direct summands we obtain the following corollary:

Corollary. *Assume that $\mathscr{V} = \bigoplus \mathscr{V}_j$ and all the $\mathscr{V}_j$'s are T-invariant. Let $T_{\mathscr{V}_j}$ denote the restriction of T to $\mathscr{V}_j$; then*

(4.4.11) $$\det T = \prod_j \det T_{\mathscr{V}_j}.$$

EXERCISES FOR SECTION 4.4

ex4.4.1 Prove that if T is nonsingular, then $\det T^{-1} = (\det T)^{-1}$.

4.5 Determinant of a matrix

4.5.1 Let $A = (a_{ij}) \in \mathscr{M}(n)$. The determinant of A can be defined in several equivalent ways. Having defined the determinant of an operator, we can define $\det A$ as the determinant of the operator T_A that A defines on $\mathbb{F}^n$ by matrix multiplication. The standard definition is equivalent, done directly by the following formula, given by (4.4.6):

$$(4.5.1) \qquad \det A = \begin{vmatrix} a_{11} & \cdots & a_{1n} \\ a_{21} & \cdots & a_{2n} \\ \vdots & \cdots & \vdots \\ a_{n1} & \cdots & a_{nn} \end{vmatrix} = \sum_{\sigma \in S_n} sgn\,[\sigma] a_{1,\sigma(1)} \cdots a_{n,\sigma(n)}.$$

Having the option to use either definition is an advantage. The first definition yields properties of $\det A$ from the properties of the determinants of operators. For example: the fact that $\det(AB) = \det A \det B$ follows immediately from 4.4.3. On the other hand, the definition by (4.5.1) is sometimes readier for computation.

4.5.2   Cofactors, expansions, and inverses. For a fixed pair (i, j), the elements in the sum in (4.5.1) that have a_{ij} as a factor are those for which $\sigma(i) = j$. Their sum is

$$(4.5.2) \qquad \sum_{\sigma \in S_n, \sigma(i)=j} sgn\,[\sigma] a_{1,\sigma(1)} \cdots a_{n,\sigma(n)} = a_{ij} A_{ij}.$$

The sum, with the factor a_{ij} removed, denoted A_{ij} in (4.5.2), is called the *cofactor* at (i, j).

We leave the proof of the following lemma as an exercise.

Lemma. *With the notation above, A_{ij} is equal to $(-1)^{i+j}$ times the determinant of the $(n-1) \times (n-1)$ submatrix obtained from A by deleting the i'th row and the j'th column.*

Partitioning the sum in (4.5.1) according to the value $\sigma(i)$ for some fixed index i gives *the expansion of the determinant along its i'th row:*

$$(4.5.3) \qquad \det A = \sum_j a_{ij} A_{ij}.$$

If we consider a "mismatched" sum: $\sum_j a_{ij} A_{kj}$ for $i \neq k$, we obtain the determinant of the matrix obtained from A by replacing the k'th row by the i'th. Since this matrix has two identical rows, its determinant is zero, that is

$$(4.5.4) \qquad \text{for } i \neq k, \qquad \sum_j a_{ij} A_{kj} = 0.$$

Finally, write $\tilde{A} = \begin{bmatrix} A_{11} & \cdots & A_{n1} \\ A_{12} & \cdots & A_{n2} \\ \vdots & \cdots & \vdots \\ A_{1n} & \cdots & A_{nn} \end{bmatrix}$ and observe that $\sum_j a_{ij} A_{kj}$ is

the ik'th entry of the matrix $A\tilde{A}$ so that equtions (4.5.3) and (4.5.4) combined are equivalent to

$$(4.5.5) \qquad\qquad A\tilde{A} = \det A\, I.$$

Proposition. *If $A \in \mathcal{M}(n)$ is nonsingular, then $A^{-1} = \frac{1}{\det(A)}\tilde{A}$.*

Historically, the matrix $\tilde{A}$ was called the *adjoint* of A, but the term *adjoint* is now used mostly in the context of duality.

EXERCISES FOR SECTION 4.5

ex4.5.1 Prove Lemma 4.5.2.

ex4.5.2 Let $A \in \mathcal{M}(n, \mathbb{F})$. Prove that $\det(-A) = (-1)^n \det A$.

ex4.5.3 A matrix $A = \{a_{ij}\} \in \mathcal{M}(n)$ is *upper triangular* if $a_{ij} = 0$ when $i > j$. A is *lower triangular* if $a_{ij} = 0$ when $i < j$. Prove that if A is either upper or lower triangular then $\det A = \prod_{i=1}^{n} a_{ii}$.

ex4.5.4 The *diagonal sum* of the matrices A_j, $j = 1, \ldots, m$ is the matrix A that has the m matrices $A_1, \ldots, A_m$ along the diagonal, and zero entries everywhere else:

$$(4.5.6) \qquad\qquad A = \begin{bmatrix} A_1 & 0 & \cdots & 0 & 0 \\ 0 & A_2 & 0 & \cdots & 0 \\ 0 & 0 & A_3 & 0 & \cdots \\ \vdots & \vdots & & \ddots & \vdots \\ 0 & 0 & 0 & 0 & A_m \end{bmatrix}.$$

Prove that $\det A = \prod \det A_j$.

ex4.5.5 Let $A = \begin{bmatrix} B & C \\ 0 & D \end{bmatrix}$, where A is an $n \times n$ matrix, B, C and D are respectively $m \times m$, $m \times (n-m)$, and $(n-m) \times (n-m)$ matrices. Give two proofs that $\det A = \det B \det D$: one by applying Proposition 4.4.5, and the second directly from definition (4.5.1).

ex4.5.6 Prove: $\det A = \det(A^{Tr})$. (A^{Tr} is the transpose of A: if $A = (a_{ij})$ then $A^{Tr} = (a_{ji})$.)

ex4.5.7 Let $A \in \mathcal{M}(n, \mathbb{R})$ be skew-symmetric. Prove that if n is odd, then $\det A = 0$.

ex4.5.8 Given $a_j \in \mathbb{F}$, $j = 0, \ldots, n-1$, prove (compute the determinant) that

$$(4.5.7) \qquad \begin{vmatrix} -\lambda & 0 & \cdots & \cdots & 0 & -a_0 \\ 1 & -\lambda & \cdots & \cdots & 0 & -a_1 \\ 0 & 1 & -\lambda & \cdots & 0 & -a_2 \\ \vdots & \vdots & & \vdots & & \vdots \\ 0 & 0 & \cdots & 1 & -\lambda & -a_{n-2} \\ 0 & 0 & & 0 & 1 & -\lambda - a_{n-1} \end{vmatrix} = (-1)^n (\lambda^n + \sum a_j \lambda^j).$$

Hint: Compute the cofactors of the elements of the last column.

ex4.5.9 How can the algorithm of reduction to row-echelon form be used to compute determinants?

ex4.5.10 Let $A \in \mathcal{M}(n)$. A defines an operator on $\mathbb{F}^n$, as well as on $\mathcal{M}(n)$, both by matrix multiplication. What is the relation between the values of $\det A$ as operator in the two cases?

ex4.5.11 Prove the following properties of the trace:

1. If $A, B \in \mathcal{M}(n)$, then $\text{trace}(A + B) = \text{trace}\,A + \text{trace}\,B$.

2. If $A \in \mathcal{M}(m, n)$ and $B \in \mathcal{M}(n, m)$, then $\text{trace}\,AB = \text{trace}\,BA$.

ex4.5.12 If $A, B \in \mathcal{M}(2)$, then $(AB - BA)^2 = -\det(AB - BA)I$.

ex4.5.13 Let $A = (a_{i,j}) \in \mathcal{M}(n)$ and let $m > n/2$. Assume that $a_{i,j} = 0$ whenever both $i \le m$ and $j \le m$. Prove that $\det(A) = 0$.

ex4.5.14 The subset of $\mathcal{M}(n, \mathbb{R})$ of matrices whose entries are integers (called *inegral matrices*) is denoted $\mathcal{M}(n, \mathbb{Z})$. Prove that an integral matrix has an inverse in $\mathcal{M}(n, \mathbb{Z})$ if and only if it is *unimodular,* that is, $\det A = \pm 1$.

ex4.5.15 *The Vandermonde determinant.* Given scalars a_j, $j = 1, \ldots, n$, the Vandermonde determinant $V(a_1, \ldots, a_n)$ is defined by

$$V(a_1, \ldots, a_n) = \begin{vmatrix} 1 & a_1 & a_1^2 & \cdots & a_1^{n-1} \\ 1 & a_2 & a_2^2 & \cdots & a_2^{n-1} \\ \vdots & \vdots & \vdots & \vdots & \vdots \\ 1 & a_n & a_n^2 & \cdots & a_n^{n-1} \end{vmatrix}.$$

Use the following steps to compute $V(a_1,\dots,a_n)$. Observe that

$$V(a_1,\dots,a_n,x) = \begin{vmatrix} 1 & a_1 & a_1^2 & \dots & a_1^n \\ 1 & a_2 & a_2^2 & \dots & a_2^n \\ \vdots & \vdots & \vdots & \vdots & \vdots \\ 1 & a_n & a_n^2 & \dots & a_n^n \\ 1 & x & x^2 & \dots & x^n \end{vmatrix}$$

is a polynomial of degree n (in x).

a. Prove that $V(a_1,\dots,a_n,x) = V(a_1,\dots,a_n)\prod_{j=1}^{n}(x-a_j)$.

b. Use induction to prove: $V(a_1,\dots,a_n) = \prod_{i<j}(a_j - a_i)$.
What is the rank of $V(a_1,\dots,a_n)$?

ex4.5.16 Let α_j, $j=1,\dots,m$, be distinct and let $\mathscr{P}$ be the space of all trigonometric polynomials of the form $P(x) = \sum_{j=1}^{m} a_j e^{i\alpha_j x}$.

a. Prove that if $P \in \mathscr{P}$ has a zero of order m (that is, a point x_0 such that $P^{(l)}(x_0) = 0$ for $l = 0,\dots,m-1$), then P is identically zero.

b. For every $k \in \mathbb{N}$ there exist constants $c_{k,l}$, $l = 0,\dots,m-1$, such that if $P \in \mathscr{P}$, then $P^{(k)}(0) = \sum_{l=0}^{m-1} c_{k,l} P^{(l)}(0)$.

c. Given $\{c_l\}$, $l = 0,\dots,m-1$, there exists $P \in \mathscr{P}$ such that $P^{(l)}(0) = c_l$ for $0 \le l < m$.

Hint: $P^{(l)}(x_0) = \sum_{j=1}^{m} a_j (i\alpha_j)^l e^{i\alpha_j x_0}$.

ex4.5.17 Let $C \in \mathscr{M}(n,\mathbb{C})$ be nonsingular. Let $\mathfrak{R}C$, resp. $\mathfrak{I}C$, be the matrix whose entries are the real parts, resp. the imaginary parts, of the corresponding entries in C. Prove that for all but a finite number of values of $a \in \mathbb{R}$, the matrix $\mathfrak{R}C + a\mathfrak{I}C$ is nonsingular.

Hint: $P(x) = \det(\mathfrak{R}C + x\mathfrak{I}C)$ is a polynomial of degree $\le n$, and $P(i) \ne 0$.

ex4.5.18 Given that the matrices $B_1, B_2 \in \mathscr{M}(n;\mathbb{R})$ are similar in $\mathscr{M}(n;\mathbb{C})$, show that they are similar in $\mathscr{M}(n;\mathbb{R})$.

Hint: If $C \in \mathscr{M}(n,\mathbb{C})$ is such that $CB_1 = B_2 C$, then $B_1, B_2 \in \mathscr{M}(n;\mathbb{R})$ implies $\mathfrak{R}CB_1 = B_2\mathfrak{R}C$ and $\mathfrak{I}CB_1 = B_2\mathfrak{I}C$.

Chapter 5

Invariant Subspaces

The study of linear systems, that is, operators on a fixed vector space $\mathcal{V}$, takes full advantage of the fact that $\mathcal{L}(\mathcal{V})$ is an algebra. Polynomials in T, that is, linear operators of the form $P(T) = \sum a_j T^j$, where $P(x) = \sum a_j x^j \in \mathbb{F}[x]$, play a crucial role in the understanding of T itself. In particular they provide a way to decompose $\mathcal{V}$ into a direct sum of T-invariant subspaces (see 4.4.5 or below), on each of which the behaviour of T is relatively simple. The key to this decomposition is the *minimal polynomial* of T, and its *prime-power factorization* (see A.6.3).

The underlying field, $\mathbb{F}$, is not listed in the notation; it is assumed known. This field may or may not be *algebraically closed* (see Definition A.6.5). This property affects much of what is discussed in this chapter, and it will appear as an explicit assumption in the statements of results that depend on it.

5.1 The characteristic polynomial

5.1.1 The characteristic polynomial of an operator. Let $(\mathcal{V}, T)$ be a linear system, and let $\{v_1, \ldots, v_n\}$ be a basis for $\mathcal{V}$. Expanding the expression $D(Tv_1 - \lambda v_1, \ldots, Tv_n - \lambda v_n)$, which appears in the definition of $\det(T - \lambda)$, we see that the latter is a polynomial, in the variable λ, of degree $n = \dim \mathcal{V}$ and leading coefficient $(-1)^n$.

DEFINITION: The *characteristic polynomial* of an operator $T \in \mathcal{L}(\mathcal{V})$ is the polynomial $\chi_T(\lambda) = \det(T - \lambda)$.

By Proposition 4.4.2, $\chi_T(\lambda) = 0$ if and only if $T - \lambda$ is singular, that is, if and only if $\ker(T - \lambda) \neq \{0\}$. The zeroes of χ_T are called *eigenvalues* of T, and the set of eigenvalues of T is called the *spectrum* of T, and denoted $\sigma(T)$.

For $\lambda \in \sigma(T)$, the space $\ker(T - \lambda)$ is called the *eigenspace* of λ. Its nonzero elements (that is, the vectors $v \neq 0$ for which $Tv = \lambda v$) are the *eigenvectors of T* corresponding to the eigenvalue λ.

Theorem. *The characteristic polynomial is conjugation-invariant: if $S, T \in \mathscr{L}(\mathscr{V})$ are conjugate then $\chi_T(\lambda) = \chi_S(\lambda)$.*

PROOF: If $S = RTR^{-1}$ with $R \in \mathbf{GL}(\mathscr{V})$ then, as $\det R^{-1} = (\det R)^{-1}$,

$$\det(S - \lambda) = \det R(T - \lambda)R^{-1} = \det(T - \lambda).$$

◀

If we write $\chi_T = \sum_0^n a_j \lambda^j$, then $a_n = (-1)^n$, and $a_0 = \det T$. Each of the coefficients a_j of χ_T is conjugation invariant and in particular so is the *trace* of T, defined by $\operatorname{trace} T = (-1)^{n-1} a_{n-1}$.

5.1.2   The characteristic polynomial of a matrix.

DEFINITION: The *characteristic polynomial* of a matrix $A \in \mathscr{M}(n)$ is the polynomial $\chi_A(\lambda) = \det(A - \lambda)$.

The reader should verify that $\chi_A = \chi_{T_A}$ where T_A is the operator of left multiplication by A on $\mathbb{F}_c^n$. The *spectrum* of the matrix A is, by definition, $\sigma(T_A)$, the set of zeros of $\chi_A = \chi_{T_A}$. The following proposition and its proof are essentially repetitions of Theorem 5.1.1.

Proposition. *If $A, B \in \mathscr{M}(n)$ are similar then they have the same characteristic polynomial. In other words, χ_A is similarity invariant.*

PROOF: If $B = CAC^{-1}$, then $B - \lambda = C(A - \lambda)C^{-1}$, which implies that
$$\det(B - \lambda) = \det C \, \det(A - \lambda) \, \det(C^{-1}) = \det(A - \lambda). \quad ◀$$

Remark: χ_A is not a *complete invariant* of the similarity class of A. Matrices (or operators) that have the same characteristic polynomials need not be similar. See exercise **ex5.1.4** below.

5.1.3 Traces. The *trace* of a matrix A is defined as follows: write the characteristic polynomial $\chi_A = \sum_0^n a_j \lambda^j$, then

$$(5.1.1) \quad a_n = (-1)^n, \quad a_0 = \det A, \quad \text{and} \quad a_{n-1} = (-1)^{n-1} \sum_1^n a_{ii}.$$

The trace of A is $\text{trace}(A) = (-1)^{n-1} a_{n-1} = \sum_1^n a_{ii}$.

The trace of an operator T is defined similarly:
$\text{trace}(T) = (-1)^{n-1} a_{n-1}$ where a_{n-1} is the coefficient of λ^{n-1} in χ_T. It is the sum of the diagonal elements in any matrix that represents T (relative to any basis).

Like any part of χ_A, the trace is similarity invariant.

∗**5.1.4** The entire characteristic polynomial can be recovered from the set $\{\text{trace}(A^k)\}_{k=1}^{n-1}$.

Proposition. *For $A \in \mathcal{M}(n;\mathbb{F})$, the coefficients of the characteristic polynomial of A are polynomials in $\{\text{trace}(A^k)\}_{k=1}^{n-1}$, with coefficients in $\mathbb{F}$.*

PROOF: Let $\lambda_1, \dots, \lambda_n$ be the eigenvalues of A (that is, the zeros of χ_A) repeated according to their multiplicity, and lying perhaps in some field extension of $\mathbb{F}$, see A.6.4. Then

$$\text{trace}(A^k) = \sum \lambda_j^k = s_k(\lambda_1, \dots, \lambda_n),$$

and the proposition follows from Corollary A.6.8 in the appendix. ◀

EXERCISES FOR SECTION 5.1

ex5.1.1 If $\mathcal{W} \subset \mathcal{V}$ is T-invariant, then $\chi_T(\lambda) = \chi_{T_{\mathcal{W}}} \chi_{T_{\mathcal{V}/\mathcal{W}}}$.

ex5.1.2 Let $T \in \mathcal{L}(\mathcal{V})$ and let $\{v_j\}_{j=1}^k$ be eigenvectors of T corresponding to distinct eigenvalues $\{\lambda_j\}_{j=1}^k$. Prove that the set $\{v_j\}_{j=1}^k$ is linearly independent.
Hint: Observe that if $Tv = \lambda v$, then $T^l v = \lambda^l v$ for all $l \in \mathbb{N}$.

ex5.1.3 Let $T \in \mathcal{L}(\mathcal{V})$ and assume that $\sigma(T)$ consists of $n = \dim \mathcal{V}$ distinct points. Prove that $\chi_T(T) = 0$.

ex5.1.4 Prove: the characteristic polynomial of an upper triangular $n \times n$ matrix $A = (a_{i,j})$ is equal to $\prod_{i=1}^{n}(a_{i,i} - \lambda)$.

Let A and B be upper triangular $n \times n$ matrices with the same diagonal elements, i.e., $a_{i,i} = b_{i,i}$. Are the two necessarily similar?

ex5.1.5 Prove: the characteristic polynomial of the $n \times n$ matrix $A = (a_{i,j})$ is equal to $\prod_{i=1}^{n}(a_{i,i} - \lambda)$ plus a polynomial of degree bounded by $n - 2$.

ex5.1.6 Assuming $\mathbb{F} = \mathbb{C}$, prove that $\text{trace}(a_{i,j})$ is equal to the sum (including multiplicity) of the zeros of the characteristic polynomial of $(a_{i,j})$. In other words, if the characteristic polynomial of the matrix $(a_{i,j})$ is equal to $\prod_{j=1}^{n}(\lambda - \lambda_j)$, then $\sum \lambda_j = \sum a_{i,i}$.

5.2   Invariant subspaces

5.2.1   Let $(\mathscr{V}, T)$ be a linear system.

DEFINITION: A subspace $\mathscr{V}_1 \subset \mathscr{V}$ is *T-invariant* if $T\mathscr{V}_1 \subseteq \mathscr{V}_1$. The entire space $\mathscr{V}$ and the trivial subspace $\{0\}$ are T-invariant for every $T \in \mathscr{L}(\mathscr{V})$. These are the *trivial invariant subspaces*.

If $\mathscr{V}_1$ is T-invariant and $v \in \mathscr{V}_1$, then $T^j v \in \mathscr{V}_1$ for all j, and taking linear combinations of these we obtain that $P(T)v \in \mathscr{V}_1$ for every polynomial P. Thus, $\mathscr{V}_1$ is $P(T)$-invariant for every $P \in \mathbb{F}[x]$.

Remarks:

a. Both $\ker(T)$ and $\text{range}(T)$ are (clearly) T-invariant.

b. If $S, T \in \mathscr{L}(\mathscr{V})$ and the two commute, then $\ker(S)$ and $\text{range}(S)$ are both T-invariant. This can be seen as follows: $S(Tv) = T(Sv) = 0$ if $Sv = 0$.

For the T-invariance of $\text{range}(S)$ observe that $TS\mathscr{V} = S(T\mathscr{V}) \subset S\mathscr{V}$.

In particular, $\ker(P(T))$ and $\text{range}(P(T))$ are T-invariant for every polynomial P in $\mathbb{F}[x]$.

c. Given $v \in \mathscr{V}$, the set $\text{span}[T, v] = \{P(T)v : P \in \mathbb{F}[x]\}$ is clearly a subspace, clearly T-invariant, and clearly the smallest T-invariant subspace containing v.

5.2.2   Assume again that $T, S \in \mathscr{L}(\mathscr{V})$, and $TS = ST$, then:

a. T commutes with $P(S)$ for every polynomial P; hence $\ker(P(S))$ and $\mathrm{range}(P(S))$ are T-invariant (see 5.2.1 **b**). In particular, for every $\lambda \in \mathbb{F}$, $\ker(S - \lambda)$ is T-invariant.

b. If $\mathscr{W}$ is an S-invariant subspace, then $T\mathscr{W}$ is S-invariant. This follows from

$$ST\mathscr{W} = TS\mathscr{W} \subset T\mathscr{W}.$$

There is no claim that $\mathscr{W}$ is T-invariant (an obvious example is $S = I$, which commutes with every operator T, and for which all subspaces are invariant). Thus, kernels offer "a special situation".

c. If v is an eigenvector for S with corresponding eigenvalue λ, i.e., $v \in \ker(S - \lambda)$ (see 5.1.1), and if (the T-invariant subspace) $\ker(S - \lambda)$ *is one-dimensional*, then v is an eigenvector for T.

If $\dim \ker(S - \lambda) > 1$, T maps v onto a vector in $\ker(S - \lambda)$, which may or may not be a scalar multiple of v. Consider the example $S = I$: S commutes with every $T \in \mathscr{L}(\mathscr{V})$, $\ker(S - 1) = \mathscr{V}$, so that every vector is an eigenvector of S, and this gives no information about its image under an arbitrary T.

5.2.3   We note that each eigenvector of T spans a one-dimensional T-invariant subspace.

Recall that the *spectrum* of T, $\sigma(T)$, is the set of all the eigenvalues of T. It is the set of zeros of the characteristic polynomial of T, $\chi_T(\lambda) = \det(T - \lambda)$ (see 5.1.1).

If the underlying field $\mathbb{F}$ is algebraically closed every nonconstant polynomial has zeros in $\mathbb{F}$ and hence the spectrum of every operator $T \in \mathscr{L}(\mathscr{V})$ is nonempty.

Proposition (Spectral mapping theorem). *Let* $(\mathscr{V}, T)$ *be a linear system,* $\lambda \in \sigma(T)$, *and* $P \in \mathbb{F}[x]$. *Then*

a. $P(\lambda) \in \sigma(P(T))$.

b. *If* $\mathbb{F}$ *is algebraically closed, then*

$$\sigma(P(T)) = \{P(\lambda) : \lambda \in \sigma(T)\} = P(\sigma(T)).$$

PROOF: **a.** Let v_λ be an eigenvector for λ, i.e., $Tv_\lambda = \lambda v_\lambda$. Then $T^j v_\lambda = \lambda^j v_\lambda$ and $P(T)v_\lambda = P(\lambda)v_\lambda$.

b. Assume that $\mathbb{F}$ is algebraically closed. For $\mu \in \mathbb{F}$, denote by $c_j(\mu)$ the roots of $P(x) - \mu$, and by m_j their multiplicities, so that

$$P(x) - \mu = \prod(x - c_j(\mu))^{m_j}, \quad \text{and} \quad P(T) - \mu = \prod(T - c_j(\mu))^{m_j}.$$

Unless $c_j(\mu) \in \sigma(T)$ for some j, all the factors are invertible, and hence so is their product.   ◄

Remark: If $\mathbb{F}$ is not algebraically closed, $\sigma(P(T))$ may be strictly larger than $P(\sigma(T))$. For example, if $\mathbb{F} = \mathbb{R}$, T the rotation by $\pi/2$ on $\mathbb{R}^2$, and $P(x) = x^2$, then $\sigma(T) = \emptyset$ while $\sigma(T^2) = \{-1\}$.

∗**5.2.4**   Part **a** of the proposition can be refined as follows:

Proposition. *Let $(\mathcal{V}, T)$ be a linear system, $\lambda \in \sigma(T)$, and $P \in \mathbb{F}[x]$. Then for all $k \in \mathbb{N}$, $\ker((P(T) - P(\lambda))^k) \supset \ker((T - \lambda)^k)$.*

PROOF: $P(x) - P(\lambda)$ vanishes for $x = \lambda$, and hence it is divisible by $x - \lambda$ (see A.6.1). If $Q \in \mathbb{F}[x]$ and $P(x) - P(\lambda) = Q(x)(x - \lambda)$, then $(P(x) - P(\lambda))^k = Q^k(x - \lambda)^k$, and $(P(T) - P(\lambda))^k = Q^k(T)(T - \lambda)^k$. Hence, if $(T - \lambda)^k v = 0$, then $(P(T) - P(\lambda))^k v = Q^k(T)(T - \lambda)^k v = 0$.
                                                                            ◄

5.2.5   As mentioned above, T-invariant subspaces are $P(T)$-invariant for all polynomials P. The converse, however, is not necessarily true. A subspace $\mathcal{W}$ can be T^2-invariant and not be T-invariant.

Examples on $\mathbb{R}^2$: T, the rotation by $\pi/2$, has no nontrivial invariant subspace while every subspace is invariant under its square $T^2 = -I$. Similarly, the reflection S which maps (x,y) to (y,x) has the diagonal $\{(x,x) : x \in \mathbb{R}\}$ as the only S-invariant subspace and yet $S^2 = I$, the identity, and "everything" is S^2-invariant.

5.2.6 Theorem. *Assume that $\mathcal{V}$ is finite-dimensional and $\mathbb{F}$ algebraically closed. Then:*

a. *Every $T \in \mathcal{L}(\mathcal{V})$ has one, or more, eigenvectors.*

b. *If S and T commute, then they have a common eigenvector.*

c. *If $\mathcal{Q} \subset \mathcal{L}(\mathcal{V})$ is a set of pairwise commuting operators, then there is a vector v which is an eigenvector of every $T \in \mathcal{Q}$.*

PROOF: **a.** This is an immediate consequence of the fact that $\sigma(T)$ is nonempty. If $\lambda \in \sigma(T)$ then $\ker(T - \lambda)$ is a nonempty invariant subspace, and every vector in it is an eigenvector for T.

b. If $\lambda \in \sigma(T)$, we have seen that $\ker(T - \lambda)$ is S-invariant.
By part **a** there is a vector $v \in \ker(T - \lambda)$ (hence, an eigenvector of T) that is an eigenvector for $S_{\ker(T-\lambda)}$, and hence for S.

c. If $\mathcal{V}$ is one-dimensional, there is nothing to prove: every $v \in \mathcal{V}$ is an eigenvector for every $T \in \mathcal{L}(\mathcal{V})$. So assume dim $\mathcal{V} > 1$. Let $\mathcal{A}$ be the set of all nontrivial subspaces of $\mathcal{V}$ that are T-invariant for every $T \in \mathcal{Q}$.

Claim: $\mathcal{A}$ is nonempty. If all the operators in $\mathcal{Q}$ are scalars (scalar multiples of the identity), then every proper subspace is in $\mathcal{A}$. Otherwise take a non-scalar $T_1 \in \mathcal{Q}$, $\lambda_1 \in \sigma(T_1)$; then $\mathcal{V}_1 = \ker(T_1 - \lambda_1)$ is a nontrivial subspace that is T-invariant for every $T \in \mathcal{Q}$.

Let $\mathcal{W}$ be a minimal element in $\mathcal{A}$ (see **ex1.3.8**); we claim that $\mathcal{W}$ is one-dimensional. Otherwise, the argument just given that $\mathcal{A}$ is nonempty applies to subspaces of $\mathcal{W}$ and we obtain a nontrivial subspace $\mathcal{W}_1 \subset \mathcal{W}$ that is T-invariant for all $T \in \mathcal{Q}$.

Every nonzero vector $w \in \mathcal{W}$ is a common eigenvector of all the operators in $\mathcal{Q}$. ◀

5.2.7 Theorem. *Let $\mathcal{W} \subset \mathcal{V}$ be a subspace, and $T \in \mathcal{L}(\mathcal{V})$. The following statements are equivalent:*

a. $\mathcal{W}$ *is T-invariant;*

b. $\mathcal{W}^\perp$ *is T^*-invariant.*

PROOF: Remember that for all $w \in \mathcal{V}$ and $u^* \in \mathcal{V}^*$ we have

$$(Tw, u^*) = (w, T^*u^*).$$

Statement **a** is equivalent to:

"for all $w \in \mathcal{W}$ and $u^* \in \mathcal{W}^\perp$ the left-hand side is identically zero."
Statement **b** is equivalent to:
"for all $w \in \mathcal{W}$ and $u^* \in \mathcal{W}^\perp$ the right-hand side is identically zero."

◀

5.2.8   The fact that, when $\mathbb{F}$ is algebraically closed, every $T \in \mathcal{L}(\mathcal{V})$ has eigenvectors, applies equally to the adjoint system $(\mathcal{V}^*, T^*)$.

Let $\mathcal{V}$ be n-dimensional and let $u^* \in \mathcal{V}^*$ be an eigenvector for T^*; then $\mathcal{V}_{n-1} = [u^*]^\perp = \{v \in \mathcal{V} : (v, u^*) = 0\}$ is a T-invariant subspace of dimension $n - 1$.

Repeating the argument in $\mathcal{V}_{n-1}$ we find a T-invariant $\mathcal{V}_{n-2} \subset \mathcal{V}_{n-1}$ of dimension $n - 2$. Repeating the argument a total of $n - 1$ times we obtain:

Theorem. *Assume that $\mathbb{F}$ is algebraically closed, and let $\mathcal{V}$ be an n-dimensional vector space over $\mathbb{F}$. For every $T \in \mathcal{L}(\mathcal{V})$, there exists a* complete flag $\{\mathcal{V}_j\}$, $j = 0, \ldots, n$, *of T-invariant subspaces of $\mathcal{V}$. That means:*

$$\mathcal{V}_0 = \{0\}, \quad \mathcal{V}_n = \mathcal{V}; \qquad \mathcal{V}_{j-1} \subset \mathcal{V}_j, \quad \text{and} \quad \dim \mathcal{V}_j = j.$$

Corollary. *If the field $\mathbb{F}$ is algebraically closed, then every matrix $A \in \mathcal{M}(n; \mathbb{F})$ is similar to an upper triangular matrix.*

PROOF: Apply the theorem to the operator T of left multiplication by A on $\mathbb{F}_{\mathbf{c}}^n$. For every $j \in [1, \ldots, n]$, choose v_j in $\mathcal{V}_j \setminus \mathcal{V}_{j-1}$.

For $l \leq n$, the set $\{v_1, \ldots, v_l\}$ is a basis for the T-invariant $\mathcal{V}_l$ so that $T v_l$ is a linear combination of $\{v_1, \ldots, v_l\}$, and the matrix B corresponding to T in the basis $\{v_1, \ldots, v_n\}$ is (upper) triangular.

The matrices A and B are similar since they represent the *same operator* relative to two bases.   ◀

Observe (again) that the assumption that $\mathbb{F}$ is algebraically closed is essential. If the underlying field is $\mathbb{R}$ (which is not algebraically closed) and T is a rotation by $\pi/2$ on $\mathbb{R}^2$, T admits *no* nontrivial invariant subspaces.

EXERCISES FOR SECTION 5.2

ex5.2.1 Let $\mathcal{W} \subset \mathcal{V}$ be T-invariant, and P a polynomial. Prove:

a. $P(T)_{\mathcal{W}} = P(T_{\mathcal{W}})$.

b. $P(T)_{\mathcal{V}/\mathcal{W}} = P(T_{\mathcal{V}/\mathcal{W}})$.

ex5.2.2 Let $\mathcal{W}$ be T-invariant. Prove that $\ker(T_{\mathcal{W}}) = \ker(T) \cap \mathcal{W}$.

ex5.2.3 Prove that every upper triangular matrix is similar to a lower triangular one (and vice versa).

ex5.2.4 If $\mathcal{V}_1 \subset \mathcal{V}$ is a subspace, then the set $\{S : S \in \mathcal{L}(\mathcal{V}), S\mathcal{V}_1 \subset \mathcal{V}_1\}$ is a subalgebra of $\mathcal{L}(\mathcal{V})$.

ex5.2.5 Show that if S and T commute and v is an eigenvector for S, it need not be an eigenvector for T (so that the assumption in the final remark of 5.2.5 that $\ker(S - \lambda)$ is one-dimensional is crucial).

ex5.2.6 Prove Theorem 5.2.8 without using duality.
Hint: Start with an eigenvector u_1 of T. Set $\mathcal{U}_1 = \mathrm{span}[u_1]$.
  Let $\tilde{u}_2 \in \mathcal{V}/\mathcal{U}_1$ be an eigenvector of $T_{\mathcal{V}/\mathcal{U}_1}$, $u_2 \in \mathcal{V}$ a representative of $\tilde{u}_2$, and $\mathcal{U}_2 = \mathrm{span}[u_1, u_2]$. Verify that $\mathcal{U}_2$ is T-invariant. Let $\tilde{u}_3 \in \mathcal{V}/\mathcal{U}_2$ be an eigenvector of $T_{\mathcal{V}/\mathcal{U}_2}$, etc.

5.3 The minimal polynomial

5.3.1 The minimal polynomial for (T, v). Assume now that $\mathcal{V}$ is an n-dimensional space. Given $T \in \mathcal{L}(\mathcal{V})$ and $v \in \mathcal{V}$, let m be the first positive integer such that $\{T^j v\}_0^m$ is linearly dependent or, equivalently, that $T^m v$ is a linear combination of $\{T^j v\}_0^{m-1}$, e.g.,[1]

$$(5.3.1) \qquad\qquad T^m v = -\sum_0^{m-1} a_j T^j v.$$

Notice that the assumption that $\{T^j v\}_0^{m-1}$ is independent guarantees that $m \le n$ and that the coefficients a_j are uniquely determined.

DEFINITION: The polynomial $\mathrm{minP}_{T,v}(x) = x^m + \sum_0^{m-1} a_j x^j$, with a_j defined by (5.3.1) is called the *minimal polynomial* for (T, v).

---

[1]The minus sign is there to give the common notation: $\mathrm{minP}_{T,v}(x) = x^m + \sum_0^{m-1} a_j x^j$.

Theorem. $\mathrm{minP}_{T,v}(x)$ *is the monic polynomial P of lowest degree that satisfies* $P(T)v = 0$.

PROOF: $\{T^j v\}_0^{m-1}$ is independent.   ◄

Remark: The set $\mathfrak{N}_{T,v} = \{P \in \mathbb{F}[x] : P(T)v = 0\}$ is an ideal in $\mathbb{F}[x]$, (see A.6.1). The theorem identifies $\mathrm{minP}_{T,v}$ as its generator. In other words: given $v \in \mathcal{V}$ and $P \in \mathbb{F}[x]$, then $P(T)v = 0$ if and only if P is divisible by $\mathrm{minP}_{T,v}$

As a simple consequence of this remark, we obtain the following important proposition.

Proposition. *For* $P \in \mathbb{F}[x]$, $P(T) = 0$ *if and only if P is divisible by* $\mathrm{minP}_{T,v}$ *for every* $v \in \mathcal{V}$.

PROOF: $P(T) = 0$ if and only if $P(T)v = 0$ for every $v \in \mathcal{V}$.   ◄

For $k \geq 0$ we have $T^{m+k}v = \sum_0^{m-1} a_j T^{j+k}v$, and induction on k proves that $T^{m+k}v \in \mathrm{span}[v, \dots, T^{m-1}v]$. It follows that $\{T^j v\}_0^{m-1}$ is a basis for $\mathrm{span}[T, v]$, and $\dim \mathrm{span}[T, v] = \deg \mathrm{minP}_{T,v}$.

If P is a polynomial, $u \in \mathcal{V}$ an arbitrary vector, and $P(T)u = 0$ then $P(T)T^k u = T^k P(T)u = 0$ for all $k \in \mathbb{N}$ so that $P(T)$ is zero on $\mathrm{span}[T, u]$. In particular $\mathrm{minP}_{T,v}(T) = 0$ on $\mathrm{span}[T, v]$.

5.3.2   Cyclic vectors. A vector $v \in \mathcal{V}$ is *cyclic* for the system $(\mathcal{V}, T)$ if $\mathrm{span}[T, v] = \mathcal{V}$. Equivalently, v is cyclic for $(\mathcal{V}, T)$ if $\mathrm{minP}_{T,v}$ is a polynomial of degree n. Not every linear system admits cyclic vectors; consider $T = I$; systems that do are called *cyclic systems*.

If v is a cyclic vector for $(\mathcal{V}, T)$ and $\mathrm{minP}_{T,v}(x) = x^n + \sum_0^{n-1} a_j x^j$, then the matrix of T with respect to the basis $\mathbf{v} = \{v, Tv, \dots, T^{n-1}v\}$ has the form

(5.3.2) $$A_{T,\mathbf{v}} = \begin{bmatrix} 0 & 0 & \dots & 0 & -a_0 \\ 1 & 0 & \dots & 0 & -a_1 \\ 0 & 1 & \dots & 0 & -a_2 \\ \vdots & \vdots & & \vdots & \vdots \\ 0 & 0 & \dots & 1 & -a_{n-1} \end{bmatrix}.$$

The characteristic polynomial of T is equal to $\det(A_{T,v} - \lambda I)$ which was shown in exercise **ex4.5.8** to be equal to $(-1)^n(\lambda^n + \sum a_j \lambda^j)$. Hence

$$(5.3.3) \qquad \chi_T(\lambda) = (-1)^n \operatorname{minP}_{T,v}(\lambda).$$

This implies, in particular, that if the operator T has a cyclic vector, then $\chi_T(T) = 0$. It is a special case, and a step in the proof, of the following theorem.

Theorem (Cayley-Hamilton). $\chi_T(T) = 0$.

PROOF: We will show that χ_T is a multiple of $\operatorname{minP}_{T,u}$ for every $u \in \mathcal{V}$, and the theorem will follow from Proposition 5.3.1.

Let $u \in \mathcal{V}$, $\mathcal{U} = \operatorname{span}[T, u]$, and $\operatorname{minP}_{T,u} = \lambda^m + \sum_0^{m-1} a_j \lambda^j$. The vectors $u, Tu, \ldots, T^{m-1}u$ form a basis for $\mathcal{U}$. Complete $\{T^j u\}_0^{m-1}$ to a basis for $\mathcal{V}$ by adding appropriate vectors $w_1, \ldots, w_{n-m}$. Let A_T be the matrix of T with respect to this basis. The top left $m \times m$ submatrix of A_T is the matrix of $T_{\mathcal{U}}$, and the $(n-m) \times m$ rectangle below it has only zero entries. It follows that $\chi_T = \chi_{T_{\mathcal{U}}} Q$, where Q is the characteristic polynomial of the $(n-m) \times (n-m)$ lower right submatrix of A.

By (5.3.3) applied to $T_{\mathcal{U}}$, we have $\chi_{T_{\mathcal{U}}} = (-1)^m \operatorname{minP}_{T,u}$, so that $\chi_T = \chi_{T_{\mathcal{U}}} Q$ is divisible by $\operatorname{minP}_{T,u}$, as claimed. ◀

An alternate way to word the proof, and to prove an additional claim along the way, is to proceed by induction on the dimension of the space $\mathcal{V}$. The additional claim is:

Proposition. *Every prime factor of χ_T divides $\operatorname{minP}_{T,u}$ for some vector $u \in \mathcal{V}$.*

PROOF: We reprove the theorem and prove the proposition.

a. If $n = 1$, both claims, of the theorem and of the proposition, are obvious.

b. Assume the claims valid for all systems of dimension smaller than n. Let $u \in \mathcal{V}$, $u \neq 0$, and $\mathcal{U} = \operatorname{span}[T, u]$. If $\mathcal{U} = \mathcal{V}$, the claims are a consequence of (5.3.3) as explained above. Otherwise, both $\mathcal{U}$

and $\mathcal{V}/\mathcal{U}$ have dimension smaller than n and, by Proposition 4.4.9 applied to $T - \lambda$ (exercise **ex5.1.1**), we have $\chi_T = \chi_{T_\mathcal{U}} \chi_{T_{\mathcal{V}/\mathcal{U}}}$. By the induction hypothesis, $\chi_{T_{\mathcal{V}/\mathcal{U}}}(T_{\mathcal{V}/\mathcal{U}}) = 0$, which means that $\chi_{T_{\mathcal{V}/\mathcal{U}}}(T)$ maps $\mathcal{V}$ into $\mathcal{U}$, and since $\chi_{T_\mathcal{U}}(T)$ maps $\mathcal{U}$ to 0, we have $\chi_T(T) = 0$.

Each prime factor of χ_T is either a factor of $\chi_{T_\mathcal{U}}$ or of $\chi_{T_{\mathcal{V}/\mathcal{U}}}$ and, by the induction hypothesis, is either a factor of $\min P_{T,u}$ or of $\min P_{T_{\mathcal{V}/\mathcal{U}},\tilde{v}}$ for some $\tilde{v} = v + \mathcal{U} \in \mathcal{V}/\mathcal{U}$. In the latter case, observe that $\min P_{T,v}(T)v = 0$. Reducing $\mod \mathcal{U}$ gives $\min P_{T,v}(T_{\mathcal{V}/\mathcal{U}})\tilde{v} = 0$, which implies that $\min P_{T_{\mathcal{V}/\mathcal{U}},\tilde{v}}$ divides $\min P_{T,v}$.   ◀

5.3.3   We return to the matrix defined in (5.3.2). Given an arbitrary monic polynomial, $P(x) = x^n + \sum b_j x^j$, the matrix

(5.3.4)

$$
\begin{bmatrix}
0 & 0 & \cdots & 0 & -b_0 \\
1 & 0 & \cdots & 0 & -b_1 \\
0 & 1 & \cdots & 0 & -b_2 \\
\vdots & \vdots & & \vdots & \vdots \\
0 & 0 & \cdots & 1 & -b_{n-1}
\end{bmatrix}
$$

is called the *companion matrix* of the polynomial P.

If $\{u_0, \ldots, u_{n-1}\}$ is a basis for $\mathcal{V}$, and we define $S \in \mathcal{L}(\mathcal{V})$ by

$$Su_j = u_{j+1} \text{ for } j < n-1, \text{ and } Su_{n-1} = -\sum_0^{n-2} b_j u_j,$$

then u_0 is cyclic for $(\mathcal{V}, S)$, the matrix (5.3.4) is the matrix $A_{S,\mathbf{u}}$ of S with respect to the basis $\mathbf{u} = \{u_0, \ldots, u_{n-1}\}$, and $\min P_{S,u_0} = P$.

Thus, *every monic polynomial* of degree n is $\min P_{S,u}$, the minimal polynomial of some cyclic vector u in an n-dimensional system $(\mathcal{V}, S)$.

5.3.4   The minimal polynomial. Let $T \in \mathcal{L}(\mathcal{V})$. The set $\mathfrak{N}_T = \{P : P \in \mathbb{F}[x], P(T) = 0\}$ is an *ideal* in $\mathbb{F}[x]$. The monic generator[2] for $\mathfrak{N}_T$ is called the *minimal polynomial* of T and denoted $\min P_T$. To put it simply: $\min P_T$ is the monic polynomial P of least degree such that $P(T) = 0$. Similarly, the minimal polynomial $\min P_A$ of a matrix $A \in$

---

[2]See A.6.1.

$\mathscr{M}(n, \mathbb{F})$ is the monic polynomial P of least degree such that $P(A) = 0$. It is the minimal polynomial of the operator T_A.

Since the dimension of $\mathscr{L}(\mathscr{V})$ is n^2, any $n^2 + 1$ powers of T are linearly dependent. This proves that $\mathfrak{N}_T$ is nontrivial and that the degree of minP_T is at most n^2. By the Cayley-Hamilton theorem, $\chi_T \in \mathfrak{N}_T$, which means that minP_T divides χ_T and its degree is therefore no greater than n.

The operator T is called *derogatory* if the degree of minP_T is smaller than the dimension n, and *nonderogatory* if the degree is n, i.e., if $\mathrm{minP}_T = \pm\chi_T$. Similarly, a matrix A is nonderogatory if $\mathrm{minP}_A = \pm\chi_A$, and derogatory otherwise, i.e., when $\deg\chi_A > \deg\mathrm{minP}_A$.

The condition "$P(T) = 0$" is equivalent to "$P(T)v = 0$ for all $v \in \mathscr{V}$", and the condition "$P(T)v = 0$" is equivalent to "$\mathrm{minP}_{T,v}$ divides minP_T". A moment's reflection gives:

Proposition. *minP_T is the least common multiple of $\mathrm{minP}_{T,v}$ for all $v \in \mathscr{V}$.*

Invoking Proposition 5.3.2 we obtain

Corollary. *Every prime factor of χ_T is a factor of minP_T.*

We shall see later (exercise **ex5.3.8**) that there are always vectors v such that minP_T is equal to $\mathrm{minP}_{T,v}$.

5.3.5 The minimal poynomial gives much information on T and on polynomials in T.

Lemma. *Let P_1 be a polynomial. Then $P_1(T)$ is invertible if and only if P_1 is relatively prime to minP_T.*

PROOF: Denote $P = \gcd(P_1, \mathrm{minP}_T)$. If P_1 is relatively prime to minP_T then $P = 1$. By Theorem A.6.2, there exist polynomials q, q_1 such that $q_1 P_1 + q\,\mathrm{minP}_T = 1$. Substituting T for x we have $q_1(T)P_1(T) = I$, so that $P_1(T)$ is invertible, and $q_1(T)$ is its inverse.

If $P \neq 1$ we write $\mathrm{minP}_T = PQ$, so that $P(T)Q(T) = \mathrm{minP}_T(T) = 0$ and hence $\ker(P(T)) \supset \mathrm{range}(Q(T))$.

Since $\deg Q < \deg \mathrm{minP_T}$, the minimality of $\mathrm{minP_T}$ guarantees that $Q(T) \neq 0$ so that $\mathrm{range}(Q(T)) \neq \{0\}$, and since P is a factor of P_1, $\ker(P_1(T)) \supset \ker(P(T)) \supset \mathrm{range}(Q(T)) \neq \{0\}$ and $P_1(T)$ is not invertible.   ◄

Comments:

a. If $P_1(x) = x$, the lemma says that T itself is invertible if and only if $\mathrm{minP_T}(0) \neq 0$. The proof for this case reads: if $\mathrm{minP_T} = xQ(x)$, and T is invertible, then $Q(T) = 0$, contradicting the minimality of $\mathrm{minP_T}$. On the other hand, if $\mathrm{minP_T}(0) = a \neq 0$, write $R(x) = a^{-1}x^{-1}(a - \mathrm{minP_T})$ and observe that $TR(T) = I - a^{-1}\mathrm{minP_T}(T) = I$, the identity, so that $R(T) = T^{-1}$.

b. If $\mathrm{minP_T}$ is $P(x)$, then $\mathrm{minP_{T+\lambda}}$, the minimal polynomial for $T + \lambda$, is $P(x - \lambda)$. It follows that $T - \lambda$ is invertible unless $x - \lambda$ divides $\mathrm{minP_T}$, that is, unless $\mathrm{minP_T}(\lambda) = 0$.

5.3.6   The following proposition is important when $\mathbb{F}$ is not algebraically closed and $\mathrm{minP_T} = \Phi$ is irreducible, but non-linear.

Proposition. *Let* $T \in \mathscr{L}(\mathscr{V})$ *be such that* $\Phi = \mathrm{minP_T}$ *is irreducible in* $\mathbb{F}[x]$. *Then* $\mathscr{P}(T) = \{P(T) : P \in \mathbb{F}[x]\}$ *is a field.*

PROOF: If $P \in \mathbb{F}[x]$ and $P(T) \neq 0$, then $\gcd(P, \Phi) = 1$ and hence $P(T)$ is invertible. Thus, every non-zero element in $\mathscr{P}(T)$ is invertible and $\mathscr{P}(T)$ is a field.   ◄

By Corollary 5.3.4, all the prime factors of χ_T are factors of, and hence equal to, $\Phi = \mathrm{minP_T}$. It follows that χ_T is a power of Φ. If $\deg \Phi = d$ and $\chi_T = \Phi^m$ then $\dim \mathscr{V} = \deg \chi_T = md$.

$\mathscr{V}$ can now be regarded as an m-dimensional vector space $\mathscr{V}_{\mathscr{P}(T)}$ over the extended field $\mathscr{P}(T)$ by treating the action of polynomials $P(T)$ on a vector v as a *multiplication of v by the field element $P(T)$*. This defines a system $(\mathscr{V}_{\mathscr{P}(T)}, T)$ in which a subspace of $\mathscr{V}_{\mathscr{P}(T)}$ is precisely a T-invariant subspace of $\mathscr{V}$.

The subspace $\mathrm{span}[T, v]$, in $\mathscr{V}$ (over $\mathbb{F}$) becomes "the line through v in $(\mathscr{V}_{\mathscr{P}(T)})$", i.e., the set of all multiples of v by scalars from $\mathscr{P}(T)$; the statement "Every subspace of a finite-dimensional vector space (here $\mathcal{V}$ over $\mathscr{P}(T)$), has a basis." translates here to: "Every T-invariant subspace of $\mathcal{V}$ is a direct sum of cyclic subspaces, that is subspaces of the form $\text{span}[T, v]$."

EXERCISES FOR SECTION 5.3

ex5.3.1 Let $T \in \mathscr{L}(\mathcal{V})$ and $v \in \mathcal{V}$. Prove that if $u \in \text{span}[T, v]$, then $\text{minP}_{T,u}$ divides $\text{minP}_{T,v}$.

ex5.3.2 Let $\mathcal{U}$ be a T-invariant subspace of $\mathcal{V}$ and $T_{\mathcal{V}/\mathcal{U}}$ the operator induced on $\mathcal{V}/\mathcal{U}$. Let $v \in \mathcal{V}$, and let $\tilde{v}$ be its image in $\mathcal{V}/\mathcal{U}$. Prove that $\text{minP}_{T_{\mathcal{V}/\mathcal{U}},\tilde{v}}$ divides $\text{minP}_{T,v}$.

ex5.3.3 If $(\mathcal{V}, T)$ is cyclic (has a cyclic vector), then every S that commutes with T is a polynomial in T. (In other words, $\mathscr{P}(T)$ is a maximal commutative subalgebra of $\mathscr{L}(\mathcal{V})$.)

Hint: If v is cyclic, and $Sv = P(T)v$ for some polynomial P, then $S = P(T)$.

ex5.3.4 a.   Assume that $\text{minP}_{T,v} = Q_1 Q_2$, and let $u = Q_1(T)v$. Prove that $\text{minP}_{T,u} = Q_2$.

b.   Let $Q \in \mathbb{F}[x]$ and assume that $\gcd(Q, \text{minP}_{T,v}) = Q_1$. Write $w = Q(T)v$. Prove that $\text{minP}_{T,w} = Q_2$.

c.   Assume that $\text{minP}_{T,v} = P_1 P_2$, with $\gcd(P_1, P_2) = 1$. Prove
$$\text{span}[T, v] = \text{span}[T, P_1(T)v] \oplus \text{span}[T, P_2(T)v].$$

ex5.3.5 Let $v_1, v_2 \in \mathcal{V}$ and assume that minP_{T,v_1} and minP_{T,v_2} are relatively prime. Prove that $\text{minP}_{T,v_1+v_2} = \text{minP}_{T,v_1} \text{minP}_{T,v_2}$.

Hint: Write $P_j = \text{minP}_{T,v_j}$, $Q = \text{minP}_{T,v_1+v_2}$, and let q_j be polynomials such that $q_1 P_1 + q_2 P_2 = 1$. Then $Q q_2 P_2(T)(v_1 + v_2) = Q(T)(v_1) = 0$, and so $P_1 \mid Q$. Similarly $P_2 \mid Q$, hence $P_1 P_2 \mid Q$. Also, $P_1 P_2(T)(v_1 + v_2) = 0$, and $Q \mid P_1 P_2$.

ex5.3.6 Prove that every singular $T \in \mathscr{L}(\mathcal{V})$ is a zero-divisor, i.e., there exists a non-zero $S \in \mathscr{L}(\mathcal{V})$ such that $ST = TS = 0$.

Hint: The constant term in minP_T is zero.

ex5.3.7 Show that if minP_T is divisible by Φ^m, with Φ irreducible, then there exist vectors $v \in \mathcal{V}$ such that $\text{minP}_{T,v} = \Phi^m$.

ex5.3.8 Show that if a polynomial P divides minP_T, there exist vectors v such that $\text{minP}_{T,v} = P$. In particular, there exist vectors $v \in \mathcal{V}$ such that $\text{minP}_{T,v} = \text{minP}_T$.

Hint: Use the prime-power factorization[3] of $\min P_T$.

ex5.3.9 $(\mathcal{V}, T)$ is cyclic if and only if $\deg \min P_T = \dim \mathcal{V}$, i.e., if and only if T is nonderogatory: $\min P_T = \chi_T$.

ex5.3.10 If $\min P_T$ is irreducible, then $\min P_{T,v} = \min P_T$ for every $v \neq 0$ in $\mathcal{V}$.

ex5.3.11 If $\min P_T$ is irreducible then $\dim \mathcal{V}$ is divisible by $\deg \min P_T$.

Hint: Use Proposition 5.3.2.

ex5.3.12 Let $P_1, P_2 \in \mathbb{F}[x]$. Prove:

$$\ker(P_1(T)) \cap \ker(P_2(T)) = \ker(\gcd(P_1, P_2)).$$

ex5.3.13 *(Schur's lemma).* A system $\{\mathcal{W}, \mathcal{S}\}$, $\mathcal{S} \subset \mathcal{L}(\mathcal{W})$, is *minimal* if no nontrivial subspace of $\mathcal{W}$ is invariant under every $S \in \mathcal{S}$.

Assume that $\{\mathcal{W}, \mathcal{S}\}$ is minimal, and $T \in \mathcal{L}(\mathcal{W})$.

a. If T commute with every $S \in \mathcal{S}$, so does $P(T)$ for every polynomial P.

b. If T commutes with every $S \in \mathcal{S}$, then $\ker(T)$ is either $\{0\}$ or $\mathcal{W}$. That means that T is either invertible or identically zero.

c. With T as above, the minimal polynomial $\min P_T$ is irreducible.

d. If T commute with every $S \in \mathcal{S}$, and the underlying field is $\mathbb{C}$, then $T = \lambda I$.

Hint: The minimal polynomial of T must be irreducible, hence linear.

ex5.3.14 Assume that T is invertible and $\deg \min P_T = m$. Prove that

$$\min P_{T^{-1}}(x) = c x^m \min P_T(x^{-1}),$$

where $c = \min P_T(0)^{-1}$.

ex5.3.15 Let $T \in \mathcal{L}(\mathcal{V})$. Prove that $\min P_T$ vanishes at every zero of χ_T.

Hint: If $Tv = \lambda v$ then $\min P_{T,v} = x - \lambda$.

Notice that if the underlying field is algebraically closed, the prime factors of χ_T are $\{x - \lambda : \lambda \in \sigma(T)\}$ and every one of these factors is $\min P_{T,v}$, where v is the corresponding eigenvector. This is the most direct proof of Proposition 5.3.2 (when $\mathbb{F}$ is algebraically closed).

ex5.3.16 Find the characteristic, and the minimal, polynomials of the 7×7 matrix $(a_{i,j})$ defined by

$$a_{i,j} = \begin{cases} 1 & \text{if } 4 \leq j = i+1 \leq 7, \\ 0 & \text{otherwise.} \end{cases}$$

---

[3] See A.6.3.

ex5.3.17 Assume that A is a non-singular matrix and let $P(x) = \sum_0^k a_j x^j$ be its minimal polynomial. Prove that $a_0 \neq 0$ and show that P gives an efficient way to compute A^{-1}.

ex5.3.18 Let $A = \begin{bmatrix} 0 & -1 \\ 1 & 0 \end{bmatrix} \in \mathcal{M}(2; \mathbb{R})$. Show that $\mathcal{P}(A) = \{P(A) : P \in \mathbb{R}[x]\}$ is isomorphic to $\mathbb{C}$.

Hint: What is $\min P_A$?

Chapter 6

Inner-Product Spaces

6.1 Inner products

Inner-product spaces, are real or complex vector spaces endowed with an additional structure, called *inner product*. The inner product allows for the introduction of geometric notions such as *distance* and *angle*. Finite-dimensional real inner-product spaces are often called *Euclidean spaces*. Complex inner-product spaces are also called *unitary spaces*.

We shall have examples of both finite- and infinite-dimensional inner-product spaces. However the theorems we prove assume, either explicitly or implicitly, that the spaces involved are *finite-dimensional*.

6.1.1   DEFINITION: An *inner product* on a *real* vector space $\mathscr{V}$ is a symmetric, positive definite bilinear form $\langle \cdot, \cdot \rangle : \mathscr{V} \times \mathscr{V} \to \mathbb{R}$. In other words, $\langle \cdot, \cdot \rangle$ is a real-valued form satisfying

1. $\langle u, v \rangle$ is bilinear.

2. $\langle u, v \rangle = \langle v, u \rangle$.

3. $\langle u, u \rangle \geq 0$ for all $u \in \mathscr{V}$, and $\langle u, u \rangle = 0$ if and only if $u = 0$.

EXAMPLES:

a.   The classical Euclidean n-space is $\mathbb{R}^n$ (also written $\mathscr{E}^n$), in which $\langle u, v \rangle = \sum x_j y_j$, where $u = [x_1, \ldots, x_n]$ and $v = [y_1, \ldots, y_n]$.

b.   The space $C_{\mathbb{R}}([0,1])$ of all continuous real-valued functions on $[0,1]$, with the inner product: $\langle f, g \rangle = \int_0^1 f(x)g(x)dx$.

6.1.2 DEFINITION: An *inner product* on a *complex* vector space $\mathscr{V}$ is a Hermitian, positive definite, sesquilinear form $\langle \cdot , \cdot \rangle : \mathscr{V} \times \mathscr{V} \to \mathbb{C}$. In other words, $\langle \cdot , \cdot \rangle$ is a complex-valued form satisfying

1. $\langle u,v \rangle$ is sesquilinear, that is, linear in u and skew linear in v:
$$\langle \lambda u, v \rangle = \lambda \langle u, v \rangle \quad \text{and} \quad \langle u, \lambda v \rangle = \overline{\lambda} \langle u, v \rangle.$$

2. $\langle u,v \rangle$ is Hermitian, that is, $\langle u, v \rangle = \overline{\langle v, u \rangle}$.

3. $\langle u, u \rangle \geq 0$, with $\langle u, u \rangle = 0$ if and only if $u = 0$.

Notice that the sesquilinearity follows from the Hermitian symmetry combined with the assumption of linearity in the first entry.

EXAMPLES:

a. In $\mathbb{C}^n$ we define the inner product of the vectors $u = [x_1, \ldots , x_n]$ and $v = [y_1, \ldots , y_n]$ by $\langle u, v \rangle = \sum x_j \overline{y}_j$. If we consider the vectors as columns, $u = \begin{bmatrix} x_1 \\ \vdots \\ x_n \end{bmatrix}$ and $v = \begin{bmatrix} y_1 \\ \vdots \\ y_n \end{bmatrix}$, then $\langle u, v \rangle = \overline{v}^{Tr} u$ (matrix multiplication).

b. The space $C([0,1])$ of all continuous complex-valued functions on $[0,1]$. The inner product is defined by $\langle f, g \rangle = \int_0^1 f(x) \overline{g(x)} dx$.

We shall reserve the notation $\mathscr{H}$ for inner-product vector spaces, whether real or complex, and to avoid duplication we write the formulae corresponding to complex $\mathscr{H}$, remembering, however, that complex conjugation is well defined on $\mathbb{R}$ (as the identity).

The distinction between real and complex $\mathscr{H}$ will be explicit when the results depend on the underlying field being *algebraically closed*, which $\mathbb{C}$ is, and $\mathbb{R}$ is not.

6.1.3   Given an inner-product space $\mathscr{H}$, we define a *norm* on it by

(6.1.1) $$\|v\| = \sqrt{\langle v, v \rangle}.$$

Lemma (The Cauchy-Schwarz inequality).

(6.1.2) $$|\langle u, v \rangle| \leq \|u\| \|v\|.$$

PROOF: If v is a scalar multiple of u, we have equality. If v, u are not proportional, then, for $\lambda \in \mathbb{R}$,

$$0 < \langle u + \lambda v, u + \lambda v \rangle = \|u\|^2 + 2\lambda \Re \langle u, v \rangle + \lambda^2 \|v\|^2.$$

The right-hand side is a quadratic polynomial (in the variable λ) with real coefficients and no real roots, which means that its discriminant, $(\Re \langle u, v \rangle)^2 - \|u\|^2 \|v\|^2$, is negative, that is, $|\Re \langle u, v \rangle| \leq \|u\| \|v\|$. This completes the proof for real $\mathcal{H}$.

If $\mathcal{H}$ is complex: we have $|\Re \langle \zeta u, v \rangle| \leq \|u\| \|v\|$ for every $\zeta \in \mathbb{C}$ with $|\zeta| = 1$. Take ζ such that $\langle \zeta u, v \rangle = |\langle u, v \rangle|$. ◄

6.1.4 The norm has the following properties:

a. Positivity: $\|0\| = 0$; if $v \neq 0$ then $\|v\| > 0$.

b. Homogeneity: $\|av\| = |a| \|v\|$ for scalars a and vectors v.

c. The triangle inequality: $\|v + u\| \leq \|v\| + \|u\|$.

d. The parallelogram law: $\|v + u\|^2 + \|v - u\|^2 = 2(\|v\|^2 + \|u\|^2)$.

Properties *a* and *b* are obvious. Property *c* is equivalent to

$$\|v\|^2 + \|u\|^2 + 2\Re \langle v, u \rangle \leq \|v\|^2 + \|u\|^2 + 2\|v\| \|u\|,$$

which reduces to (6.1.2). The parallelogram law is obtained by "opening brackets" in the inner products that correspond to $\|v + u\|^2$ and $\|v - u\|^2$.

The first three properties are common to all norms, whether defined by an inner product or not. They imply that the norm can be viewed as length, and $\delta(u, v) = \|u - v\|$ has the properties of a *metric*.

The parallelogram law, on the other hand, is specific to, and in fact characteristic of, the norms defined by an inner product.

Proposition. *A norm defined by an inner product determines the inner product.*

The proof is left to the reader as exercises **ex6.1.14** and **ex6.1.15**.

6.1.5   Orthogonality. Let $\mathcal{H}$ be an inner-product space.

DEFINITION: The vectors v, u in $\mathcal{H}$ are said to be (mutually) *orthogonal*, denoted $v \perp u$, if $\langle v, u \rangle = 0$. Observe that, since $\langle u, v \rangle = \overline{\langle v, u \rangle}$, the relation is symmetric: $u \perp v \Longleftrightarrow v \perp u$.

The vector v is *orthogonal to a set* $A \subset \mathcal{H}$, denoted $v \perp A$, if it is orthogonal to every vector in A.

If both $v \perp A$ and $u \perp A$ then, for every $w \in A$,

$$\langle av + bu, w \rangle = a\langle v, w \rangle + b\langle u, w \rangle = 0.$$

It follows that for any set $A \subset \mathcal{H}$, the set $A^{\perp} = \{v : v \perp A\}$ is a subspace of $\mathcal{H}$. (This notation is consistent with 3.1.3; see 6.2.1 below).

Similarly, if $v \perp A$, and both w_1 and w_2 are in A, then we have

$$\langle v, aw_1 + bw_2 \rangle = \bar{a}\langle v, w_1 \rangle + \bar{b}\langle v, w_2 \rangle = 0$$

so that $v \perp (\mathrm{span}[A])$. In other words: $A^{\perp} = (\mathrm{span}[A])^{\perp}$.

A vector v is said to be *normal* if $\|v\| = 1$. A sequence $\{v_1, \ldots, v_m\}$ is *orthonormal* if

(6.1.3) $\langle v_i, v_j \rangle = \delta_{i,j}$ (i.e., 1 if $i = j$, and 0 if $i \neq j$);

that is, if the vectors v_j are normal and pairwise mutually orthogonal.

Propositions.   Let $\{u_1, \ldots, u_m\}$ be orthonormal in an n-dimensional space $\mathcal{H}$.

a. $\{u_1, \ldots, u_m\}$ *is linearly independent.*
PROOF: If $\sum a_j u_j = 0$ then $a_k = \langle \sum a_j u_j, u_k \rangle = 0$ for all $k \in [1, m]$. ◀

b.   *For every* $v \in \mathcal{H}$, *the vector* $v_1 = v - \sum_1^m \langle v, u_j \rangle u_j$ *is orthogonal to* $\mathrm{span}[u_1, \ldots, u_m]$.
PROOF: $\langle v_1, u_k \rangle = \langle v, u_k \rangle - \langle v, u_k \rangle = 0$ for every $k \in [1, m]$. If w is a vector in $\mathrm{span}[u_1, \ldots, u_m]$, $w = \sum_1^m c_k u_k$, and it follows that

$$\langle v_1, w \rangle = \sum_1^m c_k \langle v_1, u_k \rangle = 0. \qquad \blacktriangleleft$$

c.   *If* $\{u_1, \ldots, u_n\}$ *is an orthonormal basis and* $v \in \mathcal{H}$, *then*

(6.1.4) $$v = \sum_1^n \langle v, u_j \rangle u_j.$$

PROOF: If $\text{span}[u_1, \ldots, u_n] = \mathcal{H}$, then $v - \sum_1^n \langle v, u_j \rangle u_j$ is orthogonal to itself, and so $v = \sum_1^n \langle v, u_j \rangle u_j$. ◄

d. *Parseval's identity.* If $\{u_1, \ldots, u_n\}$ is an orthonormal basis for $\mathcal{H}$, then for all $v, w \in \mathcal{H}$,

$$(6.1.5) \qquad \langle v, w \rangle = \sum_1^n \langle v, u_j \rangle \overline{\langle w, u_j \rangle}.$$

PROOF:

$$\langle v, w \rangle = \langle \sum \langle v, u_j \rangle u_j, \sum \langle w, u_l \rangle u_l \rangle = \sum_{j,l} \langle v, u_j \rangle \overline{\langle w, u_l \rangle} \langle u_j, u_l \rangle$$
$$= \sum_j \langle v, u_j \rangle \overline{\langle w, u_j \rangle}$$
◄

e. *Bessel's inequality and identity.* If $\{u_1, \ldots, u_m\}$ is orthonormal and $v \in \mathcal{H}$, then

$$(6.1.6) \qquad \sum |\langle v, u_j \rangle|^2 \leq \|v\|^2.$$

If $\{u_1, \ldots, u_n\}$ is an orthonormal basis for $\mathcal{H}$, then $\|v\|^2 = \sum_1^n |\langle v, u_j \rangle|^2$.

PROOF: The identity, when $\{u_1, \ldots, u_m\}$ is a basis, is clearly a special case of (6.1.5). The inequality follows from this and the fact (see Corollary 6.1.6 below) that every orthonormal $\{u_1, \ldots, u_m\}$ can be completed to an orthonormal basis. ◄

f. *The matrix* $(a_{i,j}) = A_{T,\mathbf{v}}$ *of an operator* $T \in \mathcal{L}(\mathcal{H})$ *relative to an orthonormal basis* $\mathbf{v} = \{v_1, \ldots, v_n\}$, *is given by* $a_{i,j} = \langle Tv_j, v_i \rangle$.

PROOF: The i'th entry in the j'th column of $A_{T,\mathbf{v}}$ is the coeffcient $a_{i,j}$ of v_i in the expansion of Tv_j in the basis $\mathbf{v}$; see subsection 2.4.3. If $\mathbf{v}$ is orthonormal and $T \in \mathcal{L}(\mathcal{H})$, then, by (6.1.4), we have $a_{i,j} = \langle Tv_j, v_i \rangle$. ◄

6.1.6 Proposition. (Gram-Schmidt). *Let* $\{v_1, \ldots, v_m\}$ *be independent; then there exists an orthonormal* $\{u_1, \ldots, u_m\}$ *such that for all* $k \in [1, m]$,

$$(6.1.7) \qquad \text{span}[u_1, \ldots, u_k] = \text{span}[v_1, \ldots, v_k].$$

PROOF: (By induction on m). The independence of $\{v_1, \ldots, v_m\}$ implies that $v_1 \neq 0$. Write $u_1 = v_1/\|v_1\|$. Then u_1 is normal and (6.1.7) is satisfied for $k = 1$.

Assume that $\{u_1, \ldots, u_l\}$ is orthonormal and that (6.1.7) is satisfied for $k \leq l$. Since $v_{l+1} \notin \text{span}[\{v_1, \ldots, v_l\}]$, the vector

$$\tilde{v}_{l+1} = v_{l+1} - \sum_{j=1}^{l} \langle v_{l+1}, u_j \rangle u_j$$

is nonzero and (by Lemma 6.1.5, part **b**) is orthogonal to u_j for all $j \leq l$. We set $u_{l+1} = \tilde{v}_{l+1}/\|\tilde{v}_{l+1}\|$.   ◄

Corollary. *Every orthonormal sequence* $\{u_1, \ldots, u_m\}$ *in* $\mathcal{H}$ *can be completed to an orthonormal basis. Hence, every finite-dimensional* $\mathcal{H}$ *has an orthonormal basis.*

PROOF: Observe that $\{u_1, \ldots, u_m\}$ is independent, complete it to a basis, apply the Gram-Schmidt process and notice that it does not change the vectors u_j, $1 \leq j \leq k$.   ◄

6.1.7   If $\mathcal{W} \subset \mathcal{H}$ is an m-dimensional subspace, and $\{v_j\}_1^n$ is a basis for $\mathcal{H}$ such that $\{v_j\}_1^m$ is a basis for $\mathcal{W}$, then the basis $\{u_j\}_1^n$ obtained by the Gram-Schmidt process splits into two: $\{u_j\}_1^m \cup \{u_j\}_{m+1}^n$, where $\{u_j\}_1^m$ is an orthonormal basis for $\mathcal{W}$ and $\{u_j\}_{m+1}^n$ is an orthonormal basis for $\mathcal{W}^\perp$. This gives a direct sum (in fact, orthogonal) decomposition $\mathcal{H} = \mathcal{W} \oplus \mathcal{W}^\perp$.

The map

(6.1.8) $$\pi_{\mathcal{W}}: v \mapsto \sum_{1}^{m} \langle v, u_j \rangle u_j$$

is called the *orthogonal projection onto* $\mathcal{W}$.

The *orthogonal projection onto* $\mathcal{W}$ depends only on $\mathcal{W}$ and not on the particular basis we started from. In fact, if $v = v_1 + v_2 = u_1 + u_2$ with v_1 and u_1 in $\mathcal{W}$, and both v_2 and u_2 in $\mathcal{W}^\perp$, we have

$$v_1 - u_1 = u_2 - v_2 \in \mathcal{W} \cap \mathcal{W}^\perp,$$

which means that $v_1 - u_1 = u_2 - v_2 = 0$, that is, $v_1 = u_1$ and $u_2 = v_2$.

6.1.8 The definition of the distance $\delta(v_1, v_2)$ $(= \|v_1 - v_2\|)$ between two vectors, extends to that of the *distance between a point (v $\in \mathcal{H}$)* *and a set (E $\subset \mathcal{H}$)* by setting $\delta(v, E) = \inf_{u \in E} \delta(v, u)$.

The *distance between two sets*, E_1 and E_2 in $\mathcal{H}$, is defined by

$$(6.1.9) \qquad \delta(E_1, E_2) = \inf\{\|v_1 - v_2\| : v_j \in E_j\}.$$

The proof of the following proposition is left as exercise **ex6.1.5**.

Proposition. *Let $\mathcal{W} \subset \mathcal{H}$ be a subspace, and $v \in \mathcal{H}$. Then*

$$\delta(v, \mathcal{W}) = \|v - \pi_{\mathcal{W}} v\|.$$

In other words, $\pi_{\mathcal{W}} v$ is the vector closest to v in $\mathcal{W}$.

EXERCISES FOR SECTION 6.1

ex6.1.1 Let $\mathcal{V}$ be a finite-dimensional real or complex vector space, and $\mathbf{v} = \{v_1, \ldots, v_n\}$ a basis. Explain: "declaring $\{v_1, \ldots, v_n\}$ to be orthonormal defines an inner product on $\mathcal{V}$".
Hint: Define $\langle \cdot, \cdot \rangle_{\mathbf{v}}$ by: for $u = \sum a_j v_j$ and $w = \sum b_j v_j$, set $\langle u, w \rangle_{\mathbf{v}} = \sum a_j \bar{b}_j$.
ex6.1.2 Prove that if $\mathcal{H}$ is a complex inner-product space and $T \in \mathcal{L}(\mathcal{H})$, there exists an orthonormal basis for $\mathcal{H}$ such that the matrix of T with respect to this basis is triangular.
Hint: See Corollary 5.2.8.

ex6.1.3
a. Let $\mathcal{H}$ be a real inner-product space. The vectors v, u are mutually orthogonal if and only if $\|v + u\|^2 = \|v\|^2 + \|u\|^2$.
b. If $\mathcal{H}$ is a complex inner-product space, and $v, u \in \mathcal{H}$, then the condition $\|v + u\|^2 = \|v\|^2 + \|u\|^2$ is necessary, but not sufficient, for $v \perp u$.
Hint: Consider the case "$\langle u, v \rangle$ purely imaginary".
c. If $\mathcal{H}$ is a complex inner-product space, and $v, u \in \mathcal{H}$, the condition: For all $a, b \in \mathbb{C}$, $\|av + bu\|^2 = |a|^2 \|v\|^2 + |b|^2 \|u\|^2$ is necessary and sufficient for $v \perp u$.
d. Let $\mathcal{V}$ and $\mathcal{U}$ be subspaces of $\mathcal{H}$. Prove that $\mathcal{V} \perp \mathcal{U}$ if and only if for all $v \in \mathcal{V}$ and $u \in \mathcal{U}$, $\|v + u\|^2 = \|v\|^2 + \|u\|^2$.
e. The set $\{v_1, \ldots, v_m\}$ is orthonormal if and only if $\|\sum a_j v_j\|^2 = \sum |a_j|^2$ for all choices of scalars $a_j, j = 1, \ldots, m$. (Here $\mathcal{H}$ is either real or complex.)

ex6.1.4 Show that the map $\pi_{\mathcal{W}}$ defined in (6.1.8) is an idempotent linear operator (that is, $\pi_{\mathcal{W}}^2 = \pi_{\mathcal{W}}$) and is independent of the particular basis used in its definition.

ex6.1.5 Prove Proposition 6.1.8.

ex6.1.6 Let $E_j = v_j + \mathcal{W}_j$ be affine subspaces in $\mathcal{H}$. What is $\delta(E_1, E_2)$?

ex6.1.7 Show that the sequence $\{u_1, \dots, u_m\}$ obtained by the Gram-Schmidt procedure is essentially unique: each u_j is unique up to multiplication by a number of modulus 1.

Hint: If $\{v_1, \dots, v_m\}$ is independent, and $\mathcal{W}_k = \mathrm{span}[\{v_1, \dots, v_k\}]$, $0 \le k < m$, then u_j is $c\pi_{\mathcal{W}_{j-1}^\perp} v_j$, with $|c| = \|\pi_{\mathcal{W}_{j-1}^\perp} v_j\|^{-1}$.

ex6.1.8 Let $A, B \in \mathcal{M}(n, \mathbb{C})$. Prove, by reference to the standard inner product on $\mathbb{C}^{n^2}$, that $\langle A, B \rangle = \mathrm{trace}(A\overline{B}^{Tr})$ is an inner product on $\mathcal{M}(n, \mathbb{C})$.

ex6.1.9 Let $A \in \mathcal{M}(n; \mathbb{C})$ and assume that its rows w_j, considered as vectors in $\mathbb{C}^n$, are pairwise orthogonal. Prove that $A\overline{A}^{Tr}$ is a diagonal matrix, and conclude that $|\det A| = \prod \|w_j\|$.

ex6.1.10 Let $\{v_1, \dots, v_n\} \subset \mathbb{C}^n$ be the rows of the matrix A. Prove *Hadamard's inequality:*

$$(6.1.10) \qquad\qquad |\det A| \le \prod \|v_j\|.$$

Hint: Write $\mathcal{W}_k = \mathrm{span}[\{v_1, \dots, v_k\}]$, $k = 0, \dots, n-1$, $w_j = \pi_{\mathcal{W}_{j-1}^\perp} v_j$, and apply the previous problem.

ex6.1.11 Let $v, v', u, u' \in \mathcal{H}$. Prove

$$(6.1.11) \qquad |\langle v, u \rangle - \langle v', u' \rangle| \le \|v\|\|u - u'\| + \|u'\|\|v - v'\|.$$

(Observe that this means that the inner product $\langle v, u \rangle$ is a continuous function of v and u in the metric defined by the norm.)

ex6.1.12 The standard operator norm[1] on $\mathcal{L}(\mathcal{H})$ is defined by

$$(6.1.12) \qquad\qquad \|T\| = \max_{\|v\|=1} \|Tv\|.$$

Let $A \in \mathcal{M}(n; \mathbb{C})$ be the matrix corresponding to T with respect to some orthonormal basis and denote its columns by u_j. Prove that

$$(6.1.13) \qquad \sup\|u_j\| \le \|T\| \le \left(\sum_{j=1}^n \|u_j\|^2 \right)^{\frac{1}{2}}.$$

---

[1] See *2.6.

ex6.1.13 Let $\mathcal{V} \subset \mathcal{H}$ be a subspace and π a projection on $\mathcal{V}$ along a subspace $\mathcal{W}$. Prove that $\|\pi\| = 1$ if and only if $\mathcal{W} = \mathcal{V}^{\perp}$, that is, if π is the orthogonal projection on $\mathcal{V}$.

ex6.1.14 Prove that in a real inner-product space, the inner product is determined by the norm: (*polarization formula over* $\mathbb{R}$)

$$(6.1.14) \qquad \langle u, v \rangle = \frac{1}{4}\left(\|u+v\|^2 - \|u-v\|^2\right).$$

ex6.1.15 Prove: In a complex inner-product space, the inner product is determined by the norm; in fact (*polarization formula over* $\mathbb{C}$)

$$(6.1.15) \qquad \langle u, v \rangle = \frac{1}{4}\left(\|u+v\|^2 - \|u-v\|^2 + i\|u+iv\|^2 - i\|u-iv\|^2\right).$$

ex6.1.16 Show that the polarization formula (6.1.15) does not depend on positivity: given a sesquilinear Hermitian form ψ (on a vector space over $\mathbb{C}$), define the *Quadratic form* associated with it by: $Q(v) = \psi(v, v)$. Prove

$$(6.1.16) \qquad \psi(u, v) = \frac{1}{4}\left(Q(u+v) - Q(u-v) + iQ(u+iv) - iQ(u-iv)\right).$$

ex6.1.17 Let $R, T \in \mathcal{L}(\mathcal{H})$. If $\langle Rv, v \rangle = \langle Tv, v \rangle$ for all $v \in \mathcal{H}$, then $R = T$.
Hint: Check that $\langle Rv, u \rangle = \langle Tv, u \rangle$ for all $v, u \in \mathcal{H}$.

6.2 Duality and the adjoint

6.2.1 $\mathcal{H}$ **as its own dual.** The inner product defined in $\mathcal{H}$ associates with every vector $u \in \mathcal{H}$ the linear functional $\varphi_u \colon v \mapsto \langle v, u \rangle$. In fact every linear functional is obtained this way.

Theorem. *If φ is a linear functional on a finite-dimensional inner-product space $\mathcal{H}$, then there exists a unique $w \in \mathcal{H}$ such that for all $v \in \mathcal{H}$,*

$$(6.2.1) \qquad \varphi(v) = \varphi_w(v) = \langle v, w \rangle.$$

PROOF: Let $\{u_j\}$ be an orthonormal basis, and let $w = \sum \overline{\varphi(u_j)}u_j$. For all j, we have $\varphi(u_j) = \overline{\langle w, u_j \rangle}$. For $v \in \mathcal{H}$, $v = \sum \langle v, u_j \rangle u_j$, hence

$$\varphi(v) = \sum \langle v, u_j \rangle \varphi(u_j) = \sum \langle v, u_j \rangle \overline{\langle w, u_j \rangle}$$

and by Parseval's identity this equals $\langle v, w \rangle$.   ◀

In particular, *an orthonormal basis in $\mathscr{H}$ is its own dual basis.*

6.2.2   The adjoint of an operator. Once we identify $\mathscr{H}$ with its
dual space, the adjoint of an operator $T \in \mathscr{L}(\mathscr{H})$ is again an operator
on $\mathscr{H}$. Indeed,[2] given $u \in \mathscr{H}$, the mapping $v \mapsto \langle Tv, u \rangle$ is a linear
functional and therefore equal to $v \mapsto \langle v, w \rangle$ for some $w \in \mathscr{H}$. We
write $T^* u = w$ and check that $u \mapsto w$ is linear. In other words, T^* is a
linear operator on $\mathscr{H}$, characterized by

(6.2.2) $\langle Tv, u \rangle = \langle v, T^* u \rangle.$

Lemma. *For $T \in \mathscr{L}(\mathscr{H})$, $(T^*)^* = T$.*

PROOF: $\langle v, (T^*)^* u \rangle = \langle T^* v, u \rangle = \overline{\langle u, T^* v \rangle} = \overline{\langle Tu, v \rangle} = \langle v, Tu \rangle.$ ◀

Proposition 3.2.4 reads in the present context as

Proposition. *For $T \in \mathscr{L}(\mathscr{H})$, $\mathrm{range}(T) = (\ker(T^*))^{\perp}$.*

PROOF: Since $\langle Tx, y \rangle = \langle x, T^* y \rangle$, we have $y \perp \mathrm{range}(T)$ if, and only
if $T^* y \perp \mathscr{H}$, that is if and only if $y \in \ker(T^*)$.   ◀

6.2.3   The adjoint of a matrix.

DEFINITION: The *adjoint of a matrix* $A \in \mathscr{M}(n; \mathbb{C})$ is the matrix $A^* = \overline{A}^{\mathrm{tr}}$. The matrix A is *self-adjoint*, or *Hermitian*, if $A = A^*$, i.e., if $a_{ij} = \overline{a_{ji}}$
for all i, j.

Notice that for matrices with real entries the complex conjugation is the identity, the adjoint is the transposed matrix, and *self-adjoint*
means *symmetric*.

Let $(a_{i,j}) = A_{T,\mathbf{v}}$ be the matrix of an operator T relative to an orthonormal basis $\mathbf{v}$, and $(b_{i,j}) = A_{T^*,\mathbf{v}}$ the matrix of T^* relative to the
same basis. By Proposition **f** of 6.1.5, we have $a_{i,j} = \langle Tv_j, v_i \rangle$ and

─────────────────────────

[2]This repeats the argument of section 3.2 in the current context.

$b_{i,j} = \langle T^*v_j, v_i \rangle = \langle v_j, Tv_i \rangle = \overline{a_{j,i}}$. It follows that $A_{T^*,\mathbf{v}} = (A_{T,\mathbf{v}})^*$ or, in words: *The matrix of the adjoint is the adjoint of the matrix.*

In particular, T is self-adjoint if and only if $A_{T,\mathbf{v}}$ is self-adjoint, for every orthonormal basis $\mathbf{v}$.

EXERCISES FOR SECTION 6.2

ex6.2.1 Prove that if $T, S \in \mathscr{L}(\mathscr{H})$, then $(ST)^* = T^*S^*$.

ex6.2.2 Prove that if $T \in \mathscr{L}(\mathscr{H})$, then $\ker(T^*T) = \ker(T)$.

ex6.2.3 Prove that the characteristic polynomial χ_{T^*} is the complex conjugate of χ_T.

ex6.2.4 Show that if $Tv = \lambda v$, $T^*u = \mu u$, and $\mu \neq \bar{\lambda}$, then $\langle v, u \rangle = 0$.

ex6.2.5 Show that if $\lambda = a + bi$, $a, b \in \mathbb{R}$, then
$$\|(T - \lambda)v\|^2 = \|(T - a)v\|^2 + |b|^2\|v\|^2.$$

ex6.2.6 Rewrite the proof of Theorem 6.2.1 along the following lines: If $\ker(\varphi) = \mathscr{H}$, then $\varphi = 0$ and $u^* = 0$. If not, $\dim \ker(\varphi) = \dim \mathscr{H} - 1$ and $(\ker(\varphi))^{\perp} \neq \emptyset$. Take any nonzero $\tilde{u} \in (\ker(\varphi))^{\perp}$ and set $u^* = c\tilde{u}$ where the constant c is the one that guarantees $\langle \tilde{u}, c\tilde{u} \rangle = \varphi(\tilde{u})$, that is, $\bar{c} = \|\tilde{u}\|^{-2}\varphi(\tilde{u})$.

6.3   Self-adjoint operators

6.3.1   Recall that an operator $T \in \mathscr{L}(\mathscr{H})$ is *self-adjoint* if T^* coincides with T, that is, if $\langle Tu, v \rangle = \langle u, Tv \rangle$ for every $u, v \in \mathscr{H}$.

EXAMPLES:

a.   For $T \in \mathscr{L}(\mathscr{H})$ (real or complex), the operators TT^* and T^*T are both self-adjoint.

b.   For an arbitrary operator $T \in \mathscr{L}(\mathscr{H})$, $\mathscr{H}$ complex, the operators

(6.3.1) $$\Re T = \frac{1}{2}(T + T^*) \quad \text{and} \quad \Im T = \frac{1}{2i}(T - T^*)$$

are both self adjoint (check!). They are called the *real* and *imaginary* parts of T, respectively, and we note that $T = \Re T + i\Im T$. If $\mathscr{H}$ is *real*, then $\Re T$ is self-adjoint, but $\Im T$ is not defined.

c.   If S and T are self-adjoint then so is $S + T$. In particular, if T is self-adjoint and $\lambda \in \mathbb{R}$ then $T - \lambda$ is self-adjoint.

The structure of self-adjoint operators follows from the following observations.

Proposition. *Assume that T is self-adjoint on* $\mathcal{H}$. *Then*

a. $\sigma(T) \subset \mathbb{R}$.

b. *If* $\mathcal{W} \subset \mathcal{H}$ *is T-invariant, then so is* $\mathcal{W}^{\perp}$.

c. *If* $\mathcal{W} \subset \mathcal{H}$ *is T-invariant, then the restriction* $T_{\mathcal{W}}$ *of T to* $\mathcal{W}$ *is self-adjoint.*

PROOF:

a. If $\lambda \in \sigma(T)$ and v is a corresponding eigenvector, then

$$\lambda \|v\|^2 = \langle Tv, v \rangle = \langle v, Tv \rangle = \bar{\lambda} \|v\|^2, \quad \text{so that } \lambda = \bar{\lambda}.$$

b. If $v \in \mathcal{W}^{\perp}$, then, for every $w \in \mathcal{W}$, we have $Tw \in \mathcal{W}$ and hence $\langle Tv, w \rangle = \langle v, Tw \rangle = 0$. It follows that $Tv \in \mathcal{W}^{\perp}$.

c. The condition $\langle Tw_1, w_2 \rangle = \langle w_1, Tw_2 \rangle$ is valid when $w_j \in \mathcal{W}$, since it holds for all vectors in $\mathcal{H}$.
◀

6.3.2 The previous proposition shows that a self-adjoint operator T induces an orthogonal decomposition of $\mathcal{H}$ into T-invariant subspaces. The spectral theorem shows that this decomposition can be refined so that all of the subspaces be 1-dimensional.

Theorem (The spectral theorem for self-adjoint operators). *Let* $\mathcal{H}$ *be a finite-dimensional, inner-product space, and let* $T \in \mathcal{L}(\mathcal{H})$ *be self-adjoint. Then there is an orthonormal basis* $\{u_1, \ldots, u_n\}$ *of* $\mathcal{H}$, *each element of which is an eigenvector of T.*

PROOF: The proof is by induction on $n = \dim \mathcal{H}$. If $n = 1$, then every vector in $\mathcal{H}$ is an eigenvector of T, and the statement is obvious.

Assume that $n > 1$ and that the statement is true for $m < n$. Because $\sigma(T) \subset \mathbb{R}$, it follows that $\sigma(T)$ is nonempty for both real and complex inner-product spaces. Thus, T has an eigenvector $u_n \in \mathcal{H}$, and we may assume that $\|u_n\| = 1$.

Since $\text{span}[u_n]$ is T-invariant, it follows that $\mathcal{H}' = \text{span}[u_n]^\perp$ is also T-invariant, by part **b** of the proposition above. By part **c**, $T_{\mathcal{H}'}$ is self-adjoint, so by the induction hypothesis, there is an orthonormal basis $\{u_1, \ldots, u_{n-1}\}$ of $\mathcal{H}'$ consisting of eigenvectors of T. Then $\{u_1, \ldots, u_n\}$ is the required orthnormal basis of $\mathcal{H}$. ◀

Note that the matrix of T relative to a basis of eigenvectors is diagonal.

6.3.3 If T is an operator on an inner-product space $\mathcal{H}$ and $\lambda \in \sigma(T)$, then we denote by $\mathcal{H}_\lambda$ the kernel of $T - \lambda$. That is,

$$(6.3.2) \qquad \mathcal{H}_\lambda = \ker(T - \lambda) = \{v \in \mathcal{H} : Tv = \lambda v\}.$$

$\mathcal{H}_\lambda$ is called the *eigenspace* of T corresponding to λ. We denote by π_λ the orthogonal projection on $\mathcal{H}_\lambda$.

With this notation we can restate the spectral theorem, Theorem 6.3.2, in the following form. Some authors refer to *this* as the spectral theorem for self-adjoint operators.

Theorem. *Let $\mathcal{H}$ be an inner-product space and T a self-adjoint operator on $\mathcal{H}$. Then $\mathcal{H} = \bigoplus_{\lambda \in \sigma(T)} \mathcal{H}_\lambda$, where $\mathcal{H}_{\lambda_1} \perp \mathcal{H}_{\lambda_2}$ when $\lambda_1 \neq \lambda_2$, and*

$$(6.3.3) \qquad T = \sum_{\lambda \in \sigma(T)} \lambda \pi_\lambda.$$

The decomposition $\mathcal{H} = \bigoplus_{\lambda \in \sigma(T)} \mathcal{H}_\lambda$ is often referred to as *the spectral decomposition induced by T on $\mathcal{H}$*. The representation (6.3.3) is called the *spectral decomposition* of T.

6.3.4 If $\{u_1, \ldots, u_n\}$ is an orthonormal basis whose elements are eigenvectors for T, say $Tu_j = \lambda_j u_j$, then

$$(6.3.4) \qquad Tv = \sum \lambda_j \langle v, u_j \rangle u_j$$

for all $v \in \mathcal{H}$. Consequently, writing $a_j = \langle v, u_j \rangle$ and $v = \sum a_j u_j$,

$$(6.3.5) \quad \langle Tv, v \rangle = \sum \lambda_j |a_j|^2 \qquad \text{and} \qquad \|Tv\|^2 = \sum |\lambda_j|^2 |\langle v, u_j \rangle|^2.$$

Proposition. *Assume that T is self-adjoint, then $\|T\| = \max_{\lambda \in \sigma(T)} |\lambda|$.*

PROOF: If λ_m is an eigenvalue with maximal absolute value in $\sigma(T)$, then $\|T\| \geq \|Tu_m\| = \max_{\lambda \in \sigma(T)} |\lambda|$. Conversely, by (6.3.5),

$$\|Tv\|^2 = \sum |\lambda_j|^2 |\langle v, u_j \rangle|^2 \leq \max |\lambda_j|^2 \sum |\langle v, u_j \rangle|^2 = \max |\lambda_j|^2 \|v\|^2.$$

◀

6.3.5   Commuting self-adjoint operators.

Let T be self-adjoint, and let $\mathscr{H} = \bigoplus_{\lambda \in \sigma(T)} \mathscr{H}_\lambda$ be the spectral decomposition induced by T. If S commutes with T, then S maps each $\mathscr{H}_\lambda$ into itself. Since the subspaces $\mathscr{H}_\lambda$ are mutually orthogonal, if S is self-adjoint then so is its restriction to each $\mathscr{H}_\lambda$, and we can apply Theorem 6.3.3 to every one of these restrictions and obtain, in each $\mathscr{H}_\lambda$, an orthonormal basis made up of eigenvectors of S. Since every vector in $\mathscr{H}_\lambda$ is an eigenvector for T, we obtained an orthonormal basis, each of whose elements is an eigenvector both for T and for S. We now have the decomposition

$$\mathscr{H} = \bigoplus_{\lambda \in \sigma(T), \mu \in \sigma(S)} \mathscr{H}_{\lambda,\mu},$$

where $\mathscr{H}_{\lambda,\mu} = \ker(T - \lambda) \cap \ker(S - \mu)$.

If we denote by $\pi_{\lambda,\mu}$ the orthogonal projection onto $\mathscr{H}_{\lambda,\mu}$, then

(6.3.6) $T = \sum \lambda \pi_{\lambda,\mu}$ and $S = \sum \mu \pi_{\lambda,\mu}.$

More generally, given a set of commuting, self-adjoint operators $\{T_j\}_{j=1}^m \subset \mathscr{L}(\mathscr{H})$ and eigenvalues $\lambda_j \in \sigma(T_j)$, we set

$$\mathscr{H}_{\lambda_1,\dots,\lambda_m} = \bigcap_{j=1}^m \ker(T_j - \lambda_j).$$

The *joint spectrum* of $\{T_j\}_{j=1}^m$ is the set of m-tuples

$$\sigma(T_1,\dots,T_m) = \{(\lambda_1,\dots,\lambda_m) : \mathscr{H}_{\lambda_1,\dots,\lambda_m} \neq \{0\}\}.$$

If we denote by $\pi_{\lambda_1,\ldots,\lambda_m}$ the projection onto $\mathcal{H}_{\lambda_1,\ldots,\lambda_m}$, then, as above, we have $\mathcal{H} = \oplus \mathcal{H}_{\lambda_1,\ldots,\lambda_m}$ and

(6.3.7) $\qquad T_j = \sum_{(\lambda_1,\ldots,\lambda_m) \in \sigma(T_1,\ldots,T_m)} \lambda_j \pi_{\lambda_1,\ldots,\lambda_m}, \qquad$ for $1 \leq j \leq m$.

The spaces $\mathcal{H}_{\lambda_1,\ldots,\lambda_m}$ are clearly invariant under every operator in the algebra generated by $\{T_j\}_{j=1}^m$, that is, operators of the form $P(T_1,\ldots,T_m)$, where P is a polynomial in m variables. These observations yield the following generalization of Theorem 6.3.3.

Theorem. *Let $\mathcal{H}$ be a finite-dimensional inner-product space, and let $\{T_j\}_{j=1}^m$ be commuting self-adjoint operators on $\mathcal{H}$. Let $\mathcal{A} \subset \mathcal{L}(\mathcal{H})$ be the subalgebra generated by $\{T_j\}_{j=1}^m$. Then*

(6.3.8) $\qquad \mathcal{H} = \bigoplus_{(\lambda_1,\ldots,\lambda_m) \in \sigma(T_1,\ldots,T_m)} \mathcal{H}_{\lambda_1,\ldots,\lambda_m},$

where $\mathcal{H}_{\lambda_1,\ldots,\lambda_m} \perp \mathcal{H}_{\mu_1,\ldots,\mu_m}$ if $(\lambda_1,\ldots,\lambda_m) \neq (\mu_1,\ldots,\mu_m)$, and every $S \in \mathcal{A}$ is a scalar operator on each $\mathcal{H}_{\lambda_1,\ldots,\lambda_m}$. To be more precise, if $S = P(T_1,\ldots,T_m)$, then

(6.3.9) $\qquad S = \sum_{(\lambda_1,\ldots,\lambda_m) \in \sigma(T_1,\ldots,T_m)} P(\lambda_1,\ldots,\lambda_m) \pi_{\lambda_1,\ldots,\lambda_m}.$

The verification of equation (6.3.9) is left as exercise **ex6.3.3**.

Every vector in $\mathcal{H}_{\lambda_1,\ldots,\lambda_m}$ is a common eigenvector of all the operators $S \in \mathcal{A}$. If we choose an orthonormal basis in every $\mathcal{H}_{\lambda_1,\ldots,\lambda_m}$, the union of these is an orthonormal basis of $\mathcal{H}$ with respect to which the matrices of all the operators in $\mathcal{A}$ are diagonal.

6.3.6 If $\mathcal{H}$ is *complex*, then a subalgebra $\mathcal{A} \subset \mathcal{L}(\mathcal{H})$ is *self-adjoint* if $S \in \mathcal{A}$ implies that $S^* \in \mathcal{A}$. By considering real and imaginary parts, we see that a self-adjoint algebra is generated (in fact spanned) by the self-adjoint elements it contains.

With this terminology, Theorem 6.3.5 may be restated as follows.

Theorem. *If $\mathscr{H}$ is a complex inner-product space, and $\mathscr{A}$ is a commutative, self-adjoint subalgebra of $\mathscr{L}(\mathscr{H})$, then there is an orthonormal basis $\{u_1, \dots, u_n\}$ of $\mathscr{H}$ such that every u_k is an eigenvector of every $T \in \mathscr{A}$.*

EXERCISES FOR SECTION 6.3

ex6.3.1 $T \in \mathscr{L}(\mathscr{H})$ is self-adjoint if and only if the quadratic form $Q(v) = \langle Tv, v \rangle$ is real-valued.

ex6.3.2 Let $T, S \in \mathscr{L}(\mathscr{H})$ be commuting, self-adjoint operators. Show that $P(S, T)$ is a self-adjoint operator for every polynomial P with real coefficients.

ex6.3.3 Verify equation (6.3.9).

ex6.3.4 Let $T \in \mathscr{L}(\mathscr{H})$ be self-adjoint, let $\lambda_1 \le \lambda_2 \le \cdots \le \lambda_n$ be its eigenvalues and $\{u_j\}$ the corresponding orthonormal eigenvectors. Prove the "min-max principle":

$$(6.3.10) \qquad \lambda_l = \min_{\dim \mathscr{W}=l} \max_{v \in \mathscr{W}, \|v\|=1} \langle Tv, v \rangle.$$

Hint: Every l-dimensional subspace intersects $\mathrm{span}[\{u_j\}_{j=l}^n]$ see 1.3.7.

ex6.3.5 Let $\mathscr{W} \subset \mathscr{H}$ be a subspace, and $\pi_\mathscr{W}$ the orthogonal projection onto $\mathscr{W}$. Prove that if T is self-adjoint on $\mathscr{H}$, then $\pi_\mathscr{W} T$ is self-adjoint on $\mathscr{W}$.

ex6.3.6 Let $A \in \mathscr{M}(n, \mathbb{R})$ be symmetric. Prove that χ_A has only real roots.

ex6.3.7 Assume that $A \in \mathscr{M}(m, \mathbb{C})$ is Hermitian. Let $\mathbf{z}$ denote the column vector with complex entries $z_1, \dots, z_m$, and assume that $\langle A\mathbf{z}, \mathbf{z} \rangle \ge 0$ for all choices of z_j. Prove that all the eigenvalues of A are nonnegative.

Show that if $\langle A\mathbf{z}, \mathbf{z} \rangle > 0$ for all $\mathbf{z} \ne 0$, then $\det A > 0$.

ex6.3.8 Find an operator $T \in \mathscr{L}(\mathbb{R}^2)$ that commutes with its adjoint, but has *no* eigenvectors. How does this example relate to Theorem 6.3.6?
Hint: Consider rotations.

ex6.3.9 Show that an algebra over $\mathbb{C}$ is self-adjoint if and only if it is generated by self-adjoint operators.

ex6.3.10 Let $\mathscr{B}$ be a commutative self-adjoint subalgebra of $\mathscr{L}(\mathscr{H})$. Prove that:

a. The dimension of $\mathscr{B}$ (over $\mathbb{C}$) is bounded by $\dim \mathscr{H}$.

b. $\mathscr{B}$ is generated by a single self-adjoint operator, i.e., there is an operator $T \in \mathscr{B}$ such that $\mathscr{B} = \{P(T) : P \in \mathbb{C}[x]\}$.

c. $\mathscr{B}$ is contained in a commutative self-adjoint subalgebra of $\mathscr{L}(\mathscr{H})$ of dimension $\dim \mathscr{H}$.

ex6.3.11 The Gram determinant $\det \Gamma$ of the vectors v_j, $j = 1, \ldots, m$, in $\mathscr{H}$ is the detrminant of the matrix

$$\Gamma = \Gamma(v_1, \ldots, v_m) = \begin{bmatrix} \langle v_1, v_1 \rangle & \langle v_1, v_2 \rangle & \cdots & \langle v_1, v_m \rangle \\ \langle v_2, v_1 \rangle & \langle v_2, v_2 \rangle & \cdots & \langle v_2, v_m \rangle \\ \cdots & \cdots & \cdots & \cdots \\ \langle v_m, v_1 \rangle & \langle v_m, v_2 \rangle & \cdots & \langle v_m, v_m \rangle \end{bmatrix}.$$

a. Identify the inequality $\Gamma(v_1, v_2) \geq 0$.

b. Prove that $\det \Gamma \geq 0$, and that it vanishes if and only if the vectors v_j are linearly dependent.
Hint: Let $\mathbf{z}$ denote, as above, the column vector with entries $z_1, \ldots, z_m$. Show that

$$(6.3.11) \qquad \langle \Gamma \mathbf{x}, \mathbf{x} \rangle = \left\| \sum_1^m x_j v_j \right\|^2.$$

6.4 Normal operators

We assume in this section that $\mathscr{H}$ is a complex inner-product space.

6.4.1 DEFINITION: An operator $T \in \mathscr{L}(\mathscr{H})$ is *normal* if it commutes with its adjoint, i.e., if $TT^* = T^*T$.

Self-adjoint operators are clearly normal. If $T \in \mathscr{L}(\mathscr{H})$ is normal, then $S = TT^* = T^*T$ is self-adjoint.

We observe that T is normal if and only if its real and imaginary parts, $\Re T$ and $\Im T$, commute (see example **b** of 6.3.1). Furthermore, exercise **ex6.4.3** below implies that if S and T are both normal and commute, then $\Im S$, $\Re S$, $\Im T$ and $\Re T$ all commute.

These observations and exercise **ex6.3.9** together imply the following lemma.

Lemma. *If* $\{T_1, \ldots, T_m\}$ *are commuting normal operators, then the algebra that they generate is contained in a commutative, self-adjoint algebra.*

Theorem 6.3.6 therefore implies the spectral theorem for normal operators.

Theorem (The spectral theorem for normal operators). *Let* $\mathcal{H}$ *be a finite-dimensional, complex inner-product space and* $\mathcal{A} \subset \mathcal{L}(\mathcal{H})$ *a commutative subalgebra generated by normal operators. Then there exists an orthonormal basis* $\{u_k\}$ *of* $\mathcal{H}$ *such that every* u_k *is an eigenvector for every* $T \in \mathcal{A}$.

In particular, If $T \in \mathcal{L}(\mathcal{H})$ *is normal, then* $\mathcal{H}$ *has an orthonormal basis of eigenvectors of* T.

EXERCISES FOR SECTION 6.4

ex6.4.1 Prove without using the spectral theorems:

a. If S is normal, then $\ker(S) = \ker(S^*)$.

b. If S is normal, then $\ker(S) = \ker(S^2)$.

c. If S is normal and $Sv = \lambda v$, then $S^* v = \bar{\lambda} v$.

d. S is normal if and only if $\|S^* v\| = \|Sv\|$ for all $v \in \mathcal{H}$.
Hint: See exercise **ex6.1.17**.

ex6.4.2 Prove: if S is normal, then a necessary and sufficient condition for an operator R to commute with S is that all the eigenspaces of S be R-invariant.

ex6.4.3 Show that if S is normal and R commutes with S, then R commutes with S^* as well.

ex6.4.4 If T is normal and $\sigma(T) \subset \mathbb{R}$, then T is self-adjoint.

ex6.4.5 Show that if S is normal, then S and S^* have the same eigenvectors with the corresponding eigenvalues complex conjugate. In particular, $\sigma(S^*) = \overline{\sigma(S)}$.

ex6.4.6 Let B be a commutative self-adjoint subalgebra of $\mathcal{L}(\mathcal{H})$. Prove:

a. The dimension of B is bounded by $\dim \mathcal{H}$.

b. B is generated by a single self-adjoint operator, i.e., there is an operator $T \in B$ such that $B = \{P(T) : P \in \mathbb{C}[x]\}$.

c. B is contained in a commutative self-adjoint subalgebra of $\mathscr{L}(\mathscr{H})$ of dimension dim $\mathscr{H}$.

6.5 Unitary and orthogonal operators

We mentioned in subsection 6.1.4 that the norm in $\mathscr{H}$ defines a metric, the *natural* metric for which the distance between the vectors v and u is given by $\delta(v,u) = \|v - u\|$.

Maps that preserve a metric are called *isometries* of the given metric. A particular class of isometries for the natural metric on $\mathscr{H}$ are the linear isometries, that is, operators $U \in \mathscr{L}(\mathscr{H})$ such that $\|Uv\| = \|v\|$ for all $v \in \mathscr{H}$. These are called *unitary operators* when $\mathscr{H}$ is complex, and *orthogonal operators* when $\mathscr{H}$ is real. The operator U is unitary if

$$\|Uv\|^2 = \langle Uv, Uv \rangle = \langle v, U^*Uv \rangle = \langle v, v \rangle.$$

It is an easy exercise to verify that polarization (see **ex6.1.15**) extends the equality $\langle v, U^*Uv \rangle = \langle v, v \rangle$ to

(6.5.1) $$\langle u, U^*Uv \rangle = \langle u, v \rangle,$$

for all $u, v \in \mathscr{V}$, which implies that $U^*U = I$. Since $\mathscr{H}$ is assumed finite-dimensional, a left inverse is an inverse, so $U^* = U^{-1}$. Observe that this implies that unitary operators are normal, and the spectral theorem for normal operators is valid for algebras of commuting unitary operators.

6.5.1 Proposition. *Let $\mathscr{H}$ be an inner-product space, $T \in \mathscr{L}(\mathscr{H})$. The following statements are equivalent:*

a. *T is unitary;*

b. *T maps some orthonormal basis onto an orthonormal basis;*

c. *T maps every orthonormal basis onto an orthonormal basis.*

PROOF: If T is unitary, then it maps orthonormal sequences to orthonormal sequences, which implies parts **b** and **c**.

Assume that both $\{u_1, \ldots, u_n\}$ and $\{Tu_1, \ldots, Tu_n\}$ are orthonormal bases. For any $v \in \mathscr{H}$, we have $v = \sum a_j u_j$ and $Tv = \sum a_j Tu_j$, and by Bessel's identity, $\|Tv\|^2 = \sum |a_j|^2 = \|v\|^2$, so T is unitary. ◀

The columns of the matrix of a unitary operator U relative to an orthonormal basis $\{v_j\}$ are the coefficient vectors of Uv_j and, by (6.1.5) (Parseval's identity), are orthonormal in $\mathbb{C}^n$ (respectively $\mathbb{R}^n$). Such matrices (with orthonormal columns) are called *unitary* when the underlying field is $\mathbb{C}$, and *orthogonal* when the field is $\mathbb{R}$.

The set $\mathscr{U}(n) \subset \mathscr{M}(n;\mathbb{C})$ of unitary $n \times n$ matrices is a group under matrix multiplication. It is caled the *unitary group*.

The set $\mathscr{O}(n) \subset \mathscr{M}(n;\mathbb{R})$ of orthogonal $n \times n$ matrices is a group under matrix multiplication. It is caled the *orthogonal group*.

6.5.2   Unitary equivalence. Recall that $A, B \in \mathscr{M}(n;\mathbb{F})$ are *similar*,[3] i.e., they represent the same operator relative to (possibly) different bases, if and only if $B = C^{-1}AC$ for some $C \in \mathbf{GL}(n,\mathbb{F})$. In other words, two matrices are similar if they are conjugate under the action of $\mathbf{GL}(n,\mathbb{F})$, where the conjugating matrix maps one basis onto another.

In the case that the vector space is a real or complex inner-product space, and if we want the conjugation to preserve geometric information, the two bases have to have the same geometry: corresponding basis elements must have the same length, and inner products between pairs of corresponding basis elements must be the same. This means that we have to restrict the choice of conjugating matrices to those that preserve the geometry, that is, to unitary matrices for a complex inner-product space, and to orthogonal matrices when the inner-product vector space is real.

DEFINITION: The matrices $A, B \in \mathscr{M}(n;\mathbb{C})$ are *unitarily equivalent* if there exists a matrix $U \in \mathscr{U}(n)$ such that $A = U^{-1}BU$.

The matrices $A, B \in \mathscr{M}(n;\mathbb{R})$ are *orthogonally equivalent* if there exists a matrix $O \in \mathscr{O}(n)$ such that $A = O^{-1}BO$.

Consider for example a matrix $A \in \mathscr{M}(n;\mathbb{C})$ that has n distinct eigenvalues. Then A has n independent eigenvectors which can be taken as a basis, and changing the standard basis to that basis conjugates A to a diagonal matrix. The matrix A is similar to a diagonal

---

[3]See Proposition 2.4.6.

matrix, but is unitarily equivalent to one only if its eigenvectors are pairwise orthogonal.

On the other hand, applying the Gram-Schmidt procedure to the basis obtained in the course of the proof of Corollary 5.2.8, one obtains the proof of the following theorem.

Theorem. *Every matrix $A \in \mathcal{M}(n;\mathbb{C})$ is unitarily equivalent to an upper triangular matrix.*

A representation $A = U^{-1}BU$, where U is unitary and B upper triangular, is called a *Schur decomposition* of A. We note that the matrices B and U are not unique (see exercise **ex6.5.6**).

6.5.3 Spectral theorem for Hermitian/symmetric matrices. Let $A \in \mathcal{M}(n;\mathbb{C})$, and let T_A denote the operator of left multiplication by A on $\mathbb{C}_{\mathbf{c}}^n$. The operator T_A is self-adjoint if and only if A is Hermitian.

In this case, Theorem 6.3.3 guarantees that $\mathbb{C}_{\mathbf{c}}^n$ has an orthonormal basis $\{v_j\}$ all of whose elements are eigenvectors for the operator of left multiplication by A.

The matrix of T_A relative to the basis $\{v_1, \ldots, v_n\}$ is diagonal, and the matrix C that affects the change of basis maps the orthonormal basis $\{v_1, \ldots, v_n\}$ onto the standard basis of $\mathbb{C}_{\mathbf{c}}^n$ and hence is unitary.

This proves the following theorem:

Theorem. *Every Hermitian matrix in $\mathcal{M}(n;\mathbb{C})$ is unitarily equivalent to a diagonal matrix.*

6.5.4 If $A \in \mathcal{M}(n;\mathbb{R})$ is symmetric, then the same reasoning applies to prove the following theorem:

Theorem. *Every symmetric matrix in $\mathcal{M}(n,\mathbb{R})$ is orthogonally equivalent to a diagonal matrix.*

6.5.5 Harmonic Analysis for finite abelian groups. Consider a finite abelian[4] group G, of order $|G|$. We write the group operation as

---

[4]The generalization of the following ideas to nonabelian groups is called *representation theory*. We give a brief survey of some of the basic elements of the representation theory for finite groups in Section 8.5.

addition. Denote by $C(G)$ the $|G|$-dimensional complex vector space of all complex-valued functions on G. Define an inner product in $C(G)$ by: for φ, $\psi \in C(G)$,

$$(6.5.2) \qquad \langle \varphi, \psi \rangle = \frac{1}{|G|} \sum_{x \in G} \varphi(x)\overline{\psi(x)}.$$

Define the operators T_y; $y \in G$ on $C(G)$ by:

$$(6.5.3) \qquad \text{for } f \in C(G), \quad T_y f(x) = f(x - y).$$

Observe that for $y \in G$, T_y is unitary and, if the order of y is m, the minimal polynomial $\text{minP}_{T_y}(z)$ is $z^m - 1$, and has m simple roots, namely the m'th roots of unity in $\mathbb{T}^*$ (the multiplicative group of complex numbers of absolute value 1). Observe also that the operators T_y commute, and the subalgebra they generate is therefore self-adjoint.

Let $\varphi \in C(G)$ be a nontrivial common eigenvector of the operators T_y, and denote by λ_y the corresponding eigenvalue of T_y. We have

$$(6.5.4) \qquad \varphi(x - y) = \lambda_y \varphi(x)$$

so that if φ is nontrivial, then $\varphi(x) \neq 0$ for all x, and since $|\lambda_y| = 1$, we have $|\varphi(x)|$ constant for all $x \in G$. We may replace φ by $\varphi(0)^{-1}\varphi$, that is, assume with no loss of generality that $\varphi(0) = 1$, and hence $|\varphi(x)| = 1$ for all x.

We now identify the eigenvalues λ_y by checking (6.5.4) for $x = y$. This gives $1 = \lambda_y \varphi(y)$ and, since $T_{-y} = T_y^{-1}$, we obtain

$$(6.5.5) \qquad \lambda_{-y} = \lambda_y^{-1} = \varphi(y),$$

and (6.5.4) becomes

$$(6.5.6) \qquad \varphi(x + y) = \varphi(x)\varphi(y).$$

This shows that the φ is a *character* of G, that is, a homomorphism of G into $\mathbb{T}^*$.

By Theorem 6.3.6 there exists an orthonormal basis $\{\gamma_j\}_{j=1}^{|G|}$ for $C(G)$ consisting of common eigenvectors of the operators T_y.

By the preceding discussion, with φ being a prototype of the γ_j's, we see that, normalizing if necessary so that $\gamma_j(0) = 1$, every γ_j is a character of G.

Lemma. *Distinct characters on G are mutually orthogonal.*

PROOF: If φ and ψ are distinct characters on G, then for every $y \in G$,

$$\langle \varphi, \psi \rangle = \sum_{x \in G} \varphi(x)\overline{\psi(x)} = \sum_{x \in G} \varphi(x-y)\overline{\psi(x-y)} = \varphi(y)^{-1}\psi(y)\langle \varphi, \psi \rangle$$

and there exists y such that $\varphi(y) \neq \psi(y)$. ◀

Corollary. *The set $\hat{G} = \{\gamma_j\}$ contains all the characters of G.*

PROOF: Since an orthonormal sequence in $C(G)$ is linearly independent, it cannot be a proper superset of the orthonormal basis $\{\gamma_j\}$. There is no room for additional characters. ◀

Since $\{\gamma_j\}$ is an orthonormal basis for $C(G)$, we obtain the *Fourier expansion* for $C(G)$.

Theorem. *Every $f \in C(G)$ has the following representation:*

(6.5.7) $$f(x) = \sum_{\gamma \in \hat{G}} \langle f, \gamma \rangle \gamma(x).$$

6.5.6 The set $\hat{G}$ is a multiplicative group, the product defined as the pointwise multiplication inherited from $C(G)$. It is commonly referred to as the *dual group* of G.

For $x \in G$ the map $\tilde{x} : \gamma \mapsto \gamma(x)$ is a homomorphism of $\hat{G}$ into $\mathbb{T}^*$, that is, a character on $\hat{G}$. Since different elements $x \in G$ give rise to different characters $\tilde{x}$, we obtain $|G| = |\hat{G}|$ characters on $\hat{G}$, which means that we obtain *all* the characters on $\hat{G}$. Direct checking that $\widetilde{x+y} = \tilde{x}\tilde{y}$ completes the proof that the map $x \mapsto \tilde{x}$ is an isomorphism of G onto $\hat{\hat{G}}$, the dual of $\hat{G}$.

EXERCISES FOR SECTION 6.5

ex6.5.1 Prove that the spectrum of a unitary operator is contained in the unit circle $\{z : |z| = 1\}$.

ex6.5.2 Prove that the set of *rows* of a unitary matrix is orthonormal.

ex6.5.3 Let $T \in \mathcal{L}(\mathcal{H})$ be invertible and assume that $\|T^j\|$ is uniformly bounded for $j \in \mathbb{Z}$. Prove that T is similar to a unitary operator.

ex6.5.4 Show that if $T \in \mathcal{L}(\mathcal{H})$ is self-adjoint and $\|T\| \leq 1$, then there exists a unitary operator U that commutes with T, such that $T = \frac{1}{2}(U + U^*)$. *Hint:* Recall that $\sigma(T) \subset [-1,1]$. For $\lambda_j \in \sigma(T)$ write $\zeta_j = \lambda_j + i\sqrt{1 - \lambda_j^2}$, so that $\lambda_j = \Re\zeta_j$ and $|\zeta_j| = 1$. Define: $Uv = \sum \zeta_j \langle v, u_j \rangle u_j$.

ex6.5.5 Given $A \in \mathbf{GL}(n, \mathbb{C})$, define an inner product $\langle \cdot, \cdot \rangle_A$ by

(6.5.8) $\langle v_1, v_2 \rangle_A = \langle Av_1, Av_2 \rangle$.

Prove that the group of operators that preserve the inner product $\langle \cdot, \cdot \rangle_A$ is the conjugate subgroup, $A^{-1} \mathcal{U}(n)A$, of $\mathcal{U}(n)$ in $\mathbf{GL}(n, \mathbb{C})$.

ex6.5.6 Let $A \in \mathcal{M}(n; \mathbb{C})$ be diagonal with distinct eigenvalues. Show that $A = U^{-1}BU$ is a Schur decomposition of A if and only if U is a permutation matrix.

$*$**ex6.5.7** Let G be a finite abelian group and $\hat{G}$ its dual. For every character $\gamma \in \hat{G}$, denote $\gamma(G) = \{\gamma(x) : x \in G\}$. Verify the following statements.

a. $\gamma(G)$ is a cyclic subgroup of $\mathbb{T}^*$.

b.   Let $\gamma_0 \in \hat{G}$ be such that $\gamma_0(G)$ is maximal, i.e., is not a proper subset of $\gamma(G)$ for any $\gamma \in \hat{G}$, and let $x_0 \in G$ be an element of smallest order such that $\gamma_0(x_0)$ is a generator for $\gamma_0(G)$. Denote by X_0 the subgroup of G generated by x_0, and let G_0 be the kernel of γ_0, i.e.,

$$G_0 = \{x \in G : \gamma_0(x) = 1\}.$$

Then $X_0 \cup G_0$ spans G, and $X_0 \cap G_0 = \{0\}$, so that $G = X_0 \oplus G_0$.

c.   If G_0 is nontrivial then G is *decomposable,* i.e., is a direct sum of proper subgroups. If G_0 is trivial then G is cyclic.

d.   Every finite abelian group is a direct sum of cyclic groups.

   This is essentially the **basis theorem** for finite abelian groups.

$*$**ex6.5.8** Let G be a finite abelian group and $\hat{G}$ its dual group.

a.   Prove that if G is cyclic then so is $\hat{G}$.

b.   If $G = G_1 \oplus G_2$. then $\hat{G} = \hat{G}_1 \oplus \hat{G}_2$.

c.   The dual group $\hat{G}$ of a finite abelian group G is isomorphic to G.

*6.6   Positive definite operators

6.6.1   An operator S is *nonnegative definite*, written $S \geq 0$, if it is *self-adjoint*,[5] and

(6.6.1) $\langle Sv, v \rangle \geq 0$

for every $v \in \mathcal{H}$. S is *positive definite*, written $S > 0$, if $S \geq 0$ and $\langle Sv, v \rangle = 0$ implies $v = 0$. We often drop the 'definite', and write simply *nonnegative* or *positive*, as the case may be.

Lemma. *A self-adjoint operator S is nonnegative definite if and only if $\sigma(S) \subset [0, \infty)$, and positive definite if and only if $\sigma(S) \subset (0, \infty)$.*

PROOF: Use the spectral decomposition $S = \sum_j \lambda_j \pi_j$, where $\{\lambda_j\} = \sigma(T)$. We have $\langle Sv, v \rangle = \sum \lambda_j \|\pi_j v\|^2$, and $\sum \|\pi_j v\|^2 = \|v\|^2$. This is clearly nonnegative for all $v \in \mathcal{H}$ if and only if $\lambda_j \geq 0$ for all j. It is strictly positive for all $v \neq 0$ if and only if every λ_j is positive.   ◄

6.6.2   Partial orders on the set of self-adjoint operators.   Let T and S be self-adjoint operators. The notions of positivity and nonnegativity define partial orders, ">" and "$\geq$", on the set of self-adjoint operators on $\mathcal{H}$. We write $T > S$ if $T - S > 0$, and $T \geq S$ if $T - S \geq 0$.

Proposition. *Let T and S be self-adjoint operators on $\mathcal{H}$, with $T \geq S$. Let $\sigma(T) = \{\lambda_j\}_{j=1}^n$ and $\sigma(S) = \{\mu_j\}_{j=1}^n$, both arranged in nondecreasing order. Then $\lambda_j \geq \mu_j$ for all j.*

PROOF: Use the minmax principle, exercise **ex6.3.4**:

$$\lambda_j = \min_{\dim \mathcal{W} = j} \max_{v \in \mathcal{W}, \|v\|=1} \langle Tv, v \rangle \geq \min_{\dim \mathcal{W} = j} \max_{v \in \mathcal{W}, \|v\|=1} \langle Sv, v \rangle = \mu_j.$$ ◄

Remark: The condition "$\lambda_j \geq \mu_j$ for $j = 1, \ldots, n$" is necessary but, even if T and S commute, *not sufficient* to prove that $T \geq S$, except

---

[5]The assumption that S is self-adjoint is supefluous—it follows from (6.6.1). See 8.2.3.

when $n = 1$. Consider for example the operators $T, S : \mathbb{C}^2 \to \mathbb{C}^2$ defined as left multiplication by the matrices

$$A_T = \begin{bmatrix} 2 & 0 \\ 0 & 4 \end{bmatrix} \quad \text{and} \quad A_S = \begin{bmatrix} 3 & 0 \\ 0 & 1 \end{bmatrix}.$$

The eigenvalues for T (in nondecreasing order) are $\{4, 2\}$, for S they are $\{3, 1\}$, yet for $T - S$ they are $\{3, -1\}$.

*6.7 Polar decomposition

6.7.1 Lemma. *A nonnegative operator S on $\mathcal{H}$ has a unique nonnegative square root.*

PROOF: We use the spectral theorem to write $S = \sum_{\lambda \in \sigma(S)} \lambda \pi_\lambda$, where π_λ is the orthogonal projection on $\mathcal{H}_\lambda$. Define $\sqrt{S} = \sum \sqrt{\lambda} \pi_\lambda$, where we take the nonnegative square roots of the (nonnegative) λ's. Then $\sqrt{S}$, being a linear combination with real coefficients of self-adjoint projections, is self-adjoint, and $(\sqrt{S})^2 = S$.

To show the uniqueness, let T be nonnegative with $T^2 = S$. Then T and S commute, and T preserves all the eigenspaces $\mathcal{H}_\lambda$ of S.

On $\ker(S)$, if $0 \in \sigma(S)$, then $T^2 = 0$ and, since T is self-adjoint, $T = 0$. On each $\mathcal{H}_\lambda$, for $\lambda > 0$, we have $S = \lambda I_\lambda$ (the identity operator on $\mathcal{H}_\lambda$) so that $T = \sqrt{\lambda} J_\lambda$, with $\sqrt{\lambda} > 0, J_\lambda$ positive, and $J_\lambda^2 = I_\lambda$. The eigenvalues of J_λ are ± 1, and the positivity of J_λ implies that they are all 1, so $J_\lambda = I_\lambda$ and $T = \sqrt{S}$.   ◀

6.7.2 Lemma. *Let $\mathcal{H}_j \subset \mathcal{H}$, $j = 1, 2$, be isomorphic subspaces. Let U_1 be a linear isometry $\mathcal{H}_1 \to \mathcal{H}_2$. Then there are unitary operators U on $\mathcal{H}$ that extend U_1.*

PROOF: Define $U = U_1$ on $\mathcal{H}_1$, while on $\mathcal{H}_1^\perp$ define U as an arbitrary linear isometry onto $\mathcal{H}_2^\perp$ (which has the same dimension). Extend by linearity to all of $\mathcal{H}$.   ◀

6.7.3 Lemma. *Let $A, B \in \mathcal{L}(\mathcal{H})$; then there exists a unitary operator U such that $B = AU$ if and only if $\|Av\| = \|Bv\|$ for all $v \in \mathcal{H}$. Furthermore, U is unique if and only if $\mathrm{range}(A) = \mathcal{H}$.*

PROOF: If U is unitary and $B = UA$, then $\|Bv\| = \|UAv\| = \|Av\|$, for all $v \in \mathscr{H}$.

Next, we note that $\ker(B) = \ker(A)$. So, if $\text{range}(A) = \mathscr{H}$, then $\text{range}(B) = \mathscr{H}$. If $\{u_1, \ldots, u_n\}$ is an orthonormal basis of $\mathscr{H}$, then $\{Au_1, \ldots, Au_n\}$ and $\{Bu_1, \ldots, Bu_n\}$ are both bases of $\mathscr{H}$. The map $U : Au_j \mapsto Bu_j$ extends to a linear isometry on all of $\mathscr{H}$, and $U = BA^{-1}$ is unique.

If $\ker(A) \neq \{0\}$, then let $\{u_1, \ldots, u_n\}$ be an orthonormal basis of $\mathscr{H}$ such that $\{u_1, \ldots, u_m\}$ is a basis for $\ker(A) = \ker(B)$. The subspace $\text{range}(A)$ is spanned by $\{Au_j\}_{m+1}^n$ and $\text{range}(B)$ is spanned by $\{Bu_j\}_{m+1}^n$. The map $U_1 : Au_j \mapsto Bu_j$ extends by linearity to an isometry of $\text{range}(A)$ onto $\text{range}(B)$. Now apply Lemma 6.7.2, and remember that $U = U_1$ on the range of A. Since U can be defined as an arbitrary linear isometry on the orthogonal complement of $\text{range}(A)$, it follows that U is not unique in this case. ◀

6.7.4 We observed in 6.3.1, that for any $T \in \mathscr{L}(\mathscr{H})$, the operators T^*T and TT^* are self-adjoint. Remember, however, that in general $T^*T \neq TT^*$ (unless T is normal).

For any $v \in \mathscr{H}$

$$\langle T^*Tv, v \rangle = \langle Tv, Tv \rangle = \|Tv\|^2 \quad \text{and} \quad \langle TT^*v, v \rangle = \|T^*v\|^2,$$

so that T^*T and TT^* are both nonnegative, and by Lemma 6.7.1, each has a nonnegative square root. We shall use the notation

$$(6.7.1) \qquad\qquad |T| = \sqrt{T^*T}.$$

Theorem (Polar decomposition). *Every operator $T \in \mathscr{L}(\mathscr{H})$ admits a representation as a product*

$$(6.7.2) \qquad\qquad T = U|T|,$$

where U is unitary, and $|T|$ is nonnegative.

PROOF: Observe that

$$\|Tv\|^2 = \langle Tv, Tv \rangle = \langle T^*Tv, v \rangle = \langle |T|^2 v, v \rangle = \langle |T|v, |T|v \rangle = \||T|v\|^2.$$

The theorem now follows from Lemma 6.7.3, with $A = |T|$ and $B = T$.

◄

When $\dim \mathscr{H} = 1$, this reduces to the polar decomposition of a complex number: $z = |z|e^{i\vartheta}$. Polar decomposition should not be confused with the *polarization formula*, (6.1.15).

Remark: Starting with T^* instead of T we obtain a unitary operator U_1 such that

(6.7.3) $T^* = U_1 |T^*| = U_1 \sqrt{TT^*}.$

Taking adjoints, we also obtain a representation of the form

(6.7.4) $T = |T^*| U_1^*.$

Notice, however, that in general $|T^*|$ and $|T|$ may be different. For example, let T be the map on $\mathbb{C}^2$ defined by $Tv_1 = v_2$ and $Tv_2 = 0$, where v_1, v_2 are the elements of an orthonormal basis. Then $T^*v_1 = 0$ and $T^*v_2 = v_1$ so that T^*T, as well as its nonnegative square root $|T|$, is the orthogonal projection onto the line of the scalar multiples of v_1. On the other hand, TT^*, as well as $|T^*|$, is the orthogonal projection onto the multiples of v_2. The matrices corresponding to the basis $\{v_1, v_2\}$ are:

$$T = \begin{bmatrix} 0 & 0 \\ 1 & 0 \end{bmatrix}, \quad T^* = \begin{bmatrix} 0 & 1 \\ 0 & 0 \end{bmatrix}, \quad T^*T = \begin{bmatrix} 1 & 0 \\ 0 & 0 \end{bmatrix}, \quad TT^* = \begin{bmatrix} 0 & 0 \\ 0 & 1 \end{bmatrix},$$

and $U = U_1 = \begin{bmatrix} 0 & 1 \\ 1 & 0 \end{bmatrix}.$

6.7.5   Given $T \in \mathscr{L}(\mathscr{H})$, the eigenvalues $\{\mu_1, \ldots, \mu_n\}$ of the nonnegative operator $|T| = \sqrt{T^*T}$, are called the *singular values* of T.

Let $\{u_1, \ldots, u_n\}$ be corresponding orthonormal eigenvectors, and denote $v_j = Uu_j$, where U is a unitary operator satisfying (6.7.2). Being the image of an orthonormal sequence by a unitary operator, $\{v_1, \ldots, v_n\}$ is orthonormal.

For all $v \in \mathscr{H}$, $v = \sum \langle v, u_j \rangle u_j$, $|T|v = \sum \mu_j \langle v, u_j \rangle u_j$, and

(6.7.5) $Tv = \sum \mu_j \langle v, u_j \rangle v_j.$

This is sometimes written[6] as

(6.7.6) $$T = \sum \mu_j u_j \otimes v_j.$$

EXERCISES FOR SECTION 6.7

ex6.7.1 Let $\{w_1, \ldots, w_n\}$ be an orthonormal basis for $\mathcal{H}$ and let T be the (weighted) shift operator on $\{w_1, \ldots, w_n\}$ defined by: $Tw_n = 0$ and for $j < n$, $Tw_j = (n-j)w_{j+1}$. Describe U and $|T|$ in (6.7.2), as well as $|T^*|$ and U_1 in (6.7.3).

ex6.7.2 An operator T is said to be *bounded below by* c, on a subspace $\mathcal{V} \subset \mathcal{H}$, written: $T \geq c$ on $\mathcal{V}$, if $\|Tv\| \geq c\|v\|$ for every $v \in \mathcal{V}$.

Assume now that $\{u_1, \ldots, u_n\}$ and $\{v_1, \ldots, v_n\}$ are orthonormal sequences, $\mu_1 \geq \mu_2 \geq \cdots \geq \mu_n > 0$, and $T \in \mathcal{L}(\mathcal{H})$ is defined by $T = \sum \mu_j u_j \otimes v_j$. Show that $\mu_j = \max\{c : \text{there exists a } j\text{-dimensional subspace on which } T \geq c\}$.

ex6.7.3 The following is another way to obtain (6.7.6) in a somewhat more general context. It is often referred to as the *singular value decomposition* of T.

Let $\mathcal{H}$ and $\mathcal{K}$ be inner-product spaces, and $T \in \mathcal{L}(\mathcal{H}, \mathcal{K})$ of rank $r > 0$. Write $\mu_1 = \|T\| = \max_{\|v\|=1} \|Tv\|$. Let $v_1 \in \mathcal{H}$ be a unit vector such that $\|Tv_1\| = \mu_1$, and write $z_1 = \mu_1^{-1} Tv_1$. Observe that $z_1 \in \mathcal{K}$ is a unit vector.

Prove that if $r = 1$, then $T = \mu_1 v_1 \otimes z_1$, that is, $Tv = \mu_1 \langle v, v_1 \rangle z_1$, for all $v \in \mathcal{H}$.

Assuming $r > 1$, let $T_1 = T_{\text{span}[v_1]^\perp}$, the restriction of T to the orthogonal complement of the span of v_1; let π_1 be the orthogonal projection of $\mathcal{K}$ onto $\text{span}[z_1]^\perp$. Write $\mu_2 = \|\pi_1 T_1\|$, let $v_2 \in \text{span}[v_1]^\perp$ be a unit vector such that $\|\pi_1 T_1 v_2\| = \mu_2$ and write $z_2 = \mu_2^{-1} \pi_1 T_1 v_2$.

Continue in this way to define μ_j, v_j and z_j recursively for $j \leq r$, so that

a. The sets $\{v_j\}_{j \leq r}$ and $\{z_j\}_{j \leq r}$ are orthonormal in $\mathcal{H}$, respectively $\mathcal{K}$.

b. For $m \leq r$, μ_m is defined by $\mu_m = \|\pi_{m-1} T_{m-1}\|$, where T_{m-1} denotes the restriction of T to $\text{span}[v_1, \ldots, v_{m-1}]^\perp$, and π_{m-1} is the orthogonal projection of $\mathcal{K}$ onto $\text{span}[z_1, \ldots, z_{m-1}]^\perp$.

c. v_m is such that $\|\pi_{m-1} T_{m-1} v_m\| = \mu_m$, and $z_m = \mu_m^{-1} \pi_{m-1} T_{m-1} v_m$.

---

[6]See ∗4.2.5.

Prove that with these choices,

$$(6.7.7) \qquad T = \sum \mu_j v_j \otimes z_j.$$

∗6.8   Contractions and unitary dilations

6.8.1   A *contraction* on $\mathcal{H}$ is an operator whose norm is ≤ 1, that is, such that $\|Tv\| \le \|v\|$ for every $v \in \mathcal{H}$. Unitary operators are contractions; orthogonal projections are contractions; the adjoint of a contraction is a contraction; the product of contractions is a contraction.

If $T \in \mathcal{L}(\mathcal{H})$ is a contraction, then both T^*T and TT^* are positive contractions and so are $I - TT^*$ and $I - T^*T$, and we set

$$(6.8.1) \qquad A = (I - TT^*)^{\frac{1}{2}} \quad \text{and} \quad B = (I - T^*T)^{\frac{1}{2}}.$$

Lemma. *Let T be a contraction and let A and B be defined by (6.8.1). Then $AT = TB$.*

PROOF: First note that $A^{2k} = (1 - TT^*)^k$ is a linear combination of powers of TT^*, while $B^{2k} = (1 - T^*T)^k$ is the same combination in which each power of TT^* is replaced by the same power of T^*T. Since $T(T^*T)^k = (TT^*)^k T$ for all $k \ge 0$, it follows that for every positive integer k, we have

$$(6.8.2) \qquad A^{2k}T = TB^{2k} \quad \text{and hence} \quad P(A^2)T = TP(B^2)$$

for every polynomial P.

Since $\sigma(T^*T) \cup \sigma(TT^*)$ is a finite subset of $[0,1]$, there is a polynomial $P(x) = \sum a_k x^k$ such that $P(x^2) = x$ on $\sigma(T^*T) \cup \sigma(TT^*)$, and hence $A = P(A^2)$ and $B = P(B^2)$. It now follows from (6.8.2) that $AT = P(A^2)T = TP(B^2) = TB$.   ◀

6.8.2   Let $\mathcal{H}_1$ be a subspace of an inner-product space $\mathcal{H}$, and denote by π_1 the orthogonal projection of $\mathcal{H}$ onto $\mathcal{H}_1$.

DEFINITION: A unitary operator U on $\mathcal{H}$ is a *unitary dilation* of a contraction $T \in \mathcal{L}(\mathcal{H}_1)$ if $T = (\pi_1 U)_{\mathcal{H}_1}$.

Theorem. *If $\dim \mathcal{H} = 2 \dim \mathcal{H}_1$, then every contraction in $\mathcal{L}(\mathcal{H}_1)$ has a unitary dilation in $\mathcal{H}$.*

PROOF: Write $\mathcal{H} = \mathcal{H}_1 \oplus \mathcal{H}_2$, with $\mathcal{H}_2 \perp \mathcal{H}_1$. Let $\mathbf{v} = \{v_1, \ldots, v_n\}$ be an orthonormal basis for $\mathcal{H}_1$ and $\mathbf{u} = \{u_1, \ldots, u_n\}$ an orthonormal basis for $\mathcal{H}_2$. Then $\mathbf{vu} = \{v_1, \ldots, v_n, u_1, \ldots, u_n\}$ is an orthonormal basis for $\mathcal{H}$.

Let T be a contraction on $\mathcal{H}_1$, and define A and B by (6.8.1). Let U be the operator on $\mathcal{H}$ whose matrix relative to the basis $\mathbf{vu}$ is $\begin{bmatrix} T & A \\ -B & T^* \end{bmatrix}$, where the operator names stand for the $n \times n$ matrices corresponding to them in $\mathcal{M}(n; \mathbb{C})$ relative to the basis $\mathbf{v}$ of $\mathcal{H}_1$.

The matrix of π_1 relative to $\mathbf{vu}$ is $\begin{bmatrix} I & 0 \\ 0 & 0 \end{bmatrix}$, so $\pi_1 U = \begin{bmatrix} T & A \\ 0 & 0 \end{bmatrix}$, and hence, $(\pi_1 U)_{\mathcal{H}_1} = T$.

To show that U is a unitary dilation of T, it remains to show that U is unitary, i.e., we need to show that $U^* = U^{-1}$. We have

$$(6.8.3) \quad UU^* = \begin{bmatrix} T & A \\ -B & T^* \end{bmatrix}\begin{bmatrix} T^* & -B \\ A & T \end{bmatrix} = \begin{bmatrix} TT^* + A^2 & AT - TB \\ T^*A - BT^* & B^2 + T^*T \end{bmatrix}.$$

The terms on the main diagonal of the product reduce to I, and the terms off the diagonal are adjoints of each other. Since $AT - TB = 0$, by Lemma 6.8.1, it follows that $UU^* = I$. ◀

6.8.3   One can state the theorem in terms of matrices rather than operators.

DEFINITION: A matrix in $\mathcal{M}(n; \mathbb{C})$ is *contractive* if it is the matrix of a contraction relative to an orthonormal basis.

Theorem. *A matrix in $\mathcal{M}(n; \mathbb{C})$ is contractive if and only if it is the top left quarter of some $2n \times 2n$ unitary matrix.*

EXERCISES FOR SECTION 6.8

ex6.8.1   Check that the "only if" part of Theorem 6.8.3 is done in the proof of Theorem 6.8.2, and prove the "if" part of the theorem.

ex6.8.2   An $m \times l$ complex matrix A is *contractive* if the map it defines in $\mathcal{L}(\mathbb{C}^l, \mathbb{C}^m)$ (with the standard bases) has norm ≤ 1.

An $n \times k$ *submatrix* of an $m \times l$ matrix A, $n < m$, $k \leq l$, is a matrix obtained from A by deleting $m - n$ of its rows and $l - k$ of its columns. Prove that if A is contractive, then every $n \times k$ submatrix of A is contractive.

ex6.8.3 Let M be an $m \times m$ unitary dilation of the contraction $T = 0 \in \mathcal{M}(n)$. Prove that $m \geq 2n$.

Chapter 7

Structure Theorems

7.1 Reducing subspaces

7.1.1 Let $(\mathscr{V}, T)$ be a linear system. A subspace $\mathscr{V}_1 \subset \mathscr{V}$ *reduces* T if it is T-invariant and has a T-invariant complement, that is, there exists a T-invariant subspace $\mathscr{V}_2$ such that $\mathscr{V} = \mathscr{V}_1 \oplus \mathscr{V}_2$.

EXAMPLE: If $\mathscr{V} = \mathscr{H}$ is an inner-product space and $T \in \mathscr{L}(\mathscr{H})$ is self-adjoint, then every T-invariant subspace is reducing since its orthogonal complement is also T-invariant. See Proposition 6.3.1 **b**

A system $(\mathscr{V}, T)$ that admits no reducing subspaces is called *irreducible*. We say also that T is *irreducible* on $\mathscr{V}$. A T-invariant subspace $\mathscr{W}$ *is irreducible* if $T_\mathscr{W}$ is irreducible.

Theorem. *Every system $(\mathscr{V}, T)$ can be decomposed into a direct sum of irreducible systems.*

PROOF: Use induction on $n = \dim \mathscr{V}$. If $n = 1$, the system is trivially irreducible. Assume the validity of the statement for $n < N$ and let $(\mathscr{V}, T)$ be of dimension N. If $(\mathscr{V}, T)$ is irreducible, the decomposition is trivial. If $(\mathscr{V}, T)$ is reducible, let $\mathscr{V} = \mathscr{V}_1 \oplus \mathscr{V}_2$ be a nontrivial decomposition with T-invariant $\mathscr{V}_j$. Then $\dim \mathscr{V}_j < N$; hence each of the systems $(\mathscr{V}_j, T_{\mathscr{V}_j})$ is completely decomposable, $\mathscr{V}_j = \bigoplus_k \mathscr{V}_{j,k}$ with every $\mathscr{V}_{j,k}$ T-invariant, and $\mathscr{V} = \bigoplus_{j,k} \mathscr{V}_{j,k}$ is a complete decomposition of $(\mathscr{V}, T)$. ◀

7.1.2 The reason for our interest in reducing subspaces and decompositions is that the operator can be analyzed separately on each direct summand of a decomposition. This reduces the analysis of the structure of a general linear system to that of the structure of irreducible

systems, and we shall see in subsections 7.1.6, 7.3.3, and in section
$*$7.5 that irreducible systems have a relatively simple structure:

a. If $(\mathcal{V}, T)$ is irreducible, then the minimal polynomial $\mathrm{minP_T}$ and
the characteristic polynomial χ_T coincide, and are equal to a power,
Φ^m, of an irreducible polynomial $\Phi \in \mathbb{F}[x]$.

b. Irreducible systems are cyclic.

As we saw in 5.3.2, if $(\mathcal{V}, T)$ is cyclic of dimension k and $v \in \mathcal{V}$
is a cyclic vector, then the matrix A_T, relative to the basis $\{T^j v\}_{j=0}^{k-1}$, is
the companion matrix of $\mathrm{minP_T}$.

7.1.3   A direct-sum decomposition of $\mathcal{V}$ into T-invariant subspaces
allows us to find a basis for $\mathcal{V}$, with respect to which the matrix of T
is particularly simple. This can be seen as follows.

Let $\dim \mathcal{V} = n$, and assume that $\mathcal{V} = \mathcal{V}_1 \oplus \mathcal{V}_2$, where $\mathcal{V}_1$ and $\mathcal{V}_2$
are T-invariant. Let $\{v_1, \ldots, v_n\}$ is a basis for $\mathcal{V}$ such that the first k
elements are a basis for $\mathcal{V}_1$ while the last $l = n - k$ elements are a basis
for $\mathcal{V}_2$.

The entries $a_{i,j}$ of the matrix A_T of T relative to this basis are zero
unless i and j are both $\le k$, or are both $> k$. This means that A_T
consists of two square blocks centered on the main diagonal. The first
is the $k \times k$ matrix of $T_{\mathcal{V}_1}$ (relative to the basis $\{v_1, \ldots, v_k\}$), and the
second is the $l \times l$ matrix of $T_{\mathcal{V}_2}$ (relative to $\{v_{k+1}, \ldots, v_n\}$).

More generally, assume that $\mathcal{V} = \bigoplus_{j=1}^{s} \mathcal{V}_j$, with T-invariant components $\mathcal{V}_j$ of dimension k_j. Let $\mathbf{v}_j = \{u_{j,i}\}_{i=1}^{k_j}$ be a basis for $\mathcal{V}_j$, then
$\mathbf{v} = \bigcup \mathbf{v}_j$ is a basis for $\mathcal{V}$, and we enumerate its elements by listing in
their order the elements of $\mathbf{v}_1$, followed by those of $\mathbf{v}_2$, etc.

If A_j is the matrix of $T_{\mathcal{V}_j}$ relative to the basis $\mathbf{v}_j$, for $j = 1, \ldots, s$,
then the matrix A_T, of T relative to the basis $\mathbf{v}$, is the *diagonal sum*[1]
of the matrices A_j. That is, A_T consists of the s matrices $A_1, \ldots, A_s$

---

[1]Also called the *direct sum* of A_j, $j = 1, \ldots, s$.

centered along the diagonal, and has zero entries everywhere else:

$$(7.1.1) \qquad A_T = \begin{bmatrix} A_1 & 0 & 0 & \cdots & 0 \\ 0 & A_2 & 0 & \cdots & 0 \\ 0 & 0 & A_3 & \cdots & 0 \\ \vdots & \vdots & \vdots & \ddots & \vdots \\ 0 & 0 & 0 & \cdots & A_s \end{bmatrix}.$$

7.1.4 Theorem 7.1.1 tells us that every linear system is a direct sum of irreducible subsystems, but does not tell us how to find them. Our program now is to show that the decomposition runs parallel to the decomposition of $\min P_T$ as a product of powers of irreducible polynomials, $\min P_T = \prod_j \Phi_j^{m_j}$. This leads to the *canonical prime-power decomposition* of $\mathscr{V}$ as a direct sum of the corresponding nullspaces, $\ker(\Phi_j^{m_j}(T))$.

Generally speaking, the kernel of an operator T is not reducing. The following theorem, gives a complete characterization of operators whose kernels are reducing.

Theorem. *Let $(\mathscr{V}, T)$ be finite-dimensional. $\ker(T)$ and $\mathrm{range}(T)$ are reducing subspaces if and only if $\ker(T) \cap \mathrm{range}(T) = \{0\}$, in which case, the only T-invariant complement of $\ker(T)$ is $\mathrm{range}(T)$, and vice versa.*

PROOF: If $\ker(T) \cap \mathrm{range}(T) = \{0\}$, then the sum $\ker(T) + \mathrm{range}(T)$ is a direct sum, so the dimension of the sum is the sum of the dimensions. By the *rank and nullity theorem* (see subsection 2.5.1), we have

$$\dim\left(\ker(T) \oplus \mathrm{range}(T)\right) = \nu(T) + \rho(T) = \dim \mathscr{V},$$

so $\mathscr{V} = \ker(T) \oplus \mathrm{range}(T)$. Since $\ker(T)$ and $\mathrm{range}(T)$ are T-invariant, it follows that both spaces are reducing.

If $\mathscr{V}_1$ is a T-invariant subspace such that $\mathscr{V} = \ker(T) \oplus \mathscr{V}_1$, then $\mathrm{range}(T) = T\mathscr{V}_1 \subseteq \mathscr{V}_1$. On the other hand,

$$\dim \mathscr{V}_1 = \dim \mathscr{V} - \dim(\ker(T)) = \dim(\mathrm{range}(T)),$$

so $\mathscr{V}_1 = \mathrm{range}(T)$, and $\ker(T) \cap \mathrm{range}(T) = \ker(T) \cap \mathscr{V}_1 = \{0\}$.

Finally, if $\mathcal{V}_2$ is a T-invariant subspace and $\mathcal{V} = \mathcal{V}_2 \oplus \text{range}(T)$, then $T\mathcal{V}_2 \subseteq \mathcal{V}_2 \cap \text{range}(T) = \{0\}$, so $\mathcal{V}_2 \subseteq \ker(T)$. One more appeal to the *rank and nullity theorem* shows that $\dim \mathcal{V}_2 = v(T)$, so $\mathcal{V}_2 = \ker(T)$ and $\ker(T) \cap \text{range}(T) = \{0\}$.   ◄

Remark: The uniqueness of the T-invariant complement for $\ker(T)$ and $\text{range}(T)$ is special. For general reducing subspaces there is no such uniqueness, as you can see by considering the identity operator, for which every subspace is reducing.

Corollary. $\ker(T)$ *reduces* T *if and only if*
$$\ker(T^2) = \ker(T).$$

PROOF: For every $T \in \mathcal{L}(\mathcal{V})$ we have $\ker(T^2) \supseteq \ker(T)$. The inclusion is proper if and only if there exist vectors v such that $Tv \neq 0$ but $T^2v = 0$, which amounts to $Tv \in \text{range}(T) \cap \ker(T)$.   ◄

7.1.5   Basic decomposition. Let $(\mathcal{V}, T)$ be an n-dimensional system. In the theorem below, we describe the fundamental relationship between factorizations of minP_T and decompositions of $\mathcal{V}$ into direct sums of T-invariant subspaces. We begin with a proposition that illuminates the basic ideas.

Proposition. *If the polynomials P_1 and P_2 are relatively prime, then*

(7.1.2) $$\ker(P_1(T)) \cap \ker(P_2(T)) = \{0\}.$$

If also $P_1(T)P_2(T) = 0$, then $\mathcal{V} = \ker(P_1(T)) \oplus \ker(P_2(T))$, and the corresponding projections are polynomials in T.

PROOF: Given that P_1 and P_2 are relatively prime, there exist, by Corollary A.6.2, polynomials q_1, q_2 such that $q_1 P_1 + q_2 P_2 = 1$. Substituting T for the variable we have

(7.1.3) $$q_1(T)P_1(T) + q_2(T)P_2(T) = I.$$

If $v \in \ker(P_1(T)) \cap \ker(P_2(T))$, that is, $P_1(T)v = P_2(T)v = 0$, we have $v = q_1(T)P_1(T)v + q_2(T)P_2(T)v = 0$. This proves (7.1.2), which implies, in particular, that $\dim \ker(P_1(T)) + \dim \ker(P_2(T)) \leq n$.

If $P_1(T)P_2(T) = 0$, then $\text{range}(P_1(T)) \subseteq \ker(P_2(T))$ and likewise, $\text{range}(P_2(T)) \subseteq \ker(P_1(T))$. By the rank and nullity theorem

(7.1.4)
$$\begin{aligned} n &= \dim \ker(P_1(T)) + \dim \text{range}(P_1(T)) \\ &\leq \dim \ker(P_1(T)) + \dim \ker(P_2(T)) \leq n. \end{aligned}$$

It follows that $\dim \ker(P_1(T)) + \dim \ker(P_2(T)) = n$, which implies that $\ker(P_1(T)) \oplus \ker(P_2(T))$ is all of $\mathscr{V}$.

Equation (7.1.3) implies that

$$\varphi_2(T) = q_1(T)P_1(T) = I - q_2(T)P_2(T)$$

is the identity on $\ker(P_2(T))$ and zero on $\ker(P_1(T))$, i.e., $\varphi_2(T)$ is the projection onto $\ker(P_2(T))$ along $\ker(P_1(T))$.

Similarly, $\varphi_1(T) = q_2(T)P_2(T)$ is the projection onto $\ker(P_1(T))$ along $\ker(P_2(T))$. ◀

Remark: The proof also shows that, if $P_1 P_2 = 0$, then

$$\text{range}(P_1(T)) = \ker(P_2(T)) \quad \text{and} \quad P_2(T) = \ker(P_1(T)).$$

The general relationship between factorizations of minP_T and decompositions of $(\mathscr{V}, T)$ is described by the following theorem.

Theorem. *For every factorization,* $\text{minP}_T = \prod_{j=1}^{l} P_j$, *into pairwise relatively prime factors, we have a direct sum decomposition of $\mathscr{V}$ into T-invariant subspaces,*

(7.1.5)
$$\mathscr{V} = \bigoplus_{j=1}^{l} \ker(P_j(T)),$$

and the projection onto each direct summand is a polynomial in T.

PROOF: We use induction on the number of factors, l.

For $l = 2$, the theorem follows immediately from the preceding proposition, so assume that $l > 2$, and that the statement is true for $l - 1$. Write $Q = \prod_{j=2}^{l} P_j$, then by the same proposition we have

$$\mathscr{V} = \ker(P_1(T)) \oplus \ker(Q(T)).$$

Applying the induction hypothesis to $(\ker(Q(T)), T_{\ker(Q(T))})$ gives

$$\ker(Q(T)) = \bigoplus_{j=2}^{l} \ker(P_j(T)),$$

and the stated decomposition of $\mathscr{V}$ follows.

For $1 \leq j \leq l$, let $Q_j = \prod_{i \neq j} P_i$; then $\gcd(P_j, Q_j) = 1$, so by Corollary A.6.2 there exist polynomials q_j and p_j such that $q_j P_j + p_j Q_j = 1$. Evaluating this identity at T gives $q_j(T)P_j(T) + p_j(T)Q_j(T) = I$, from which it follows, as it did in Proposition 7.1.5, that the polynomial

(7.1.6) $\varphi_j(T) = p_j(T)Q_j(T) = I - q_j(T)P_j(T)$

is the projection onto $\ker(P_j(T))$ along $\bigoplus_{i \neq j} \ker(P_i(T))$.   ◀

Notice that the projections $\varphi_j(T)$ all commute, since they are all polynomials in T. Also, they are *mutually orthogonal in pairs*, in the sense that $\varphi_i(T)\varphi_j(T) = 0$ when $j \neq i$.

7.1.6   The canonical prime-power decomposition. Applying Theorem 7.1.5 to the prime-power factorization of $\mathrm{minP_T}$,

$$\mathrm{minP_T} = \prod \Phi_j^{m_j},$$

where the Φ_j's are distinct irreducible (prime) polynomials in $\mathbb{F}[x]$, and m_j their respective multiplicities,[2] we obtain the *canonical prime-power decomposition* of $(\mathscr{V}, T)$:

(7.1.7) $\mathscr{V} = \bigoplus_{j=1}^{k} \ker(\Phi_j^{m_j}(T)).$

The spaces $\ker(\Phi_j^{m_j}(T))$ are called the *primary components* of $(\mathscr{V}, T)$.

The projection onto $\ker(\Phi_j^{m_j}(T))$ is a polynomial in T, as given above in (7.1.6),

(7.1.8) $\varphi_j(T) = I - q_j(T)\Phi_j^{m_j}(T) = p_j(T)\prod_{i \neq j}\Phi_i^{m_i}(T),$

---

[2]See Theorem A.6.3.

where q_j and p_j are polynomials satisfying $q_j \Phi_j^{m_j} + p_j \prod_{i \neq j} \Phi_i^{m_i} = 1$.

If $\mathscr{W} \subset \mathscr{V}$ is T-invariant, then the subspaces

$$\varphi_j(T)\mathscr{W} = \mathscr{W} \cap \ker(\Phi_j^{m_j}(T)),$$

are T-invariant and we have a decomposition

(7.1.9) $$\mathscr{W} = \bigoplus_{j=1}^{k} \varphi_j(T)\mathscr{W}.$$

Proposition. *The T-invariant subspace $\mathscr{W}$ is reducing if and only if $\varphi_j(T)\mathscr{W}$ is a reducing subspace of $\ker(\Phi_j^{m_j}(T))$ for every j.*

PROOF: If $\mathscr{W}$ is reducing and $\mathscr{U}$ is a T-invariant complement, then

$$\ker(\Phi_j^{m_j}(T)) = \varphi_j(T)\mathscr{V} = \varphi_j(T)\mathscr{W} \oplus \varphi_j(T)\mathscr{U},$$

and both components are T-invariant. Conversely, if $\mathscr{U}_j$ is T-invariant and $\ker(\Phi_j^{m_j}(T)) = \varphi_j(T)\mathscr{W} \oplus \mathscr{U}_j$, then $\mathscr{U} = \bigoplus \mathscr{U}_j$ is an invariant complement to $\mathscr{W}$. ◀

7.1.7 By the Cayley-Hamilton theorem and Corollary 5.3.4, the prime factors of χ_T are those of $\mathrm{minP_T}$, with at least the same multiplicities, that is:

(7.1.10) $$\chi_T = \prod \Phi_j^{s_j}, \quad \text{with} \quad s_j \geq m_j.$$

The minimal polynomial of the restriction of T to $\ker(\Phi_j^{m_j}(T))$ is $\Phi_j^{m_j}$ and its characteristic polynomial is $\Phi_j^{s_j}$. The dimension of $\ker(\Phi_j^{m_j}(T))$ is $s_j \deg(\Phi_j)$.

7.1.8 When the underlying field $\mathbb{F}$ is algebraically closed, and in particular when $\mathbb{F} = \mathbb{C}$, every irreducible polynomial in $\mathbb{F}[x]$ is linear and every polynomial is a product of linear factors, see A.6.6.

Recall that the *spectrum* of T is the set $\sigma(T) = \{\lambda_j\}$ of zeros of χ_T or, equivalently, of $\mathrm{minP_T}$. For systems over an algebraically closed field the prime-power factorization of $\mathrm{minP_T}$ has the form

$$\mathrm{minP_T} = \prod_{\lambda \in \sigma(T)} (x - \lambda)^{m(\lambda)}$$

where $m(\lambda)$ is the multiplicity of λ in $\mathrm{minP_T}$.

The space $\mathscr{V}_\lambda = \ker((T - \lambda)^{m(\lambda)})$ is called the *nilspace*, or the *generalized eigenspace*, of λ. The canonical decomposition of $(\mathscr{V}, T)$ is given by:

$$(7.1.11) \qquad\qquad \mathscr{V} = \bigoplus_{\lambda \in \sigma(T)} \mathscr{V}_\lambda.$$

The same decomposition obtains more generally when all the prime factors of $\mathrm{minP_T}$ are linear (even when $\mathbb{F}$ is not algebraically closed).

EXERCISES FOR SECTION 7.1

ex7.1.1 Let $T \in \mathscr{L}(\mathscr{V})$, $k > 0$ and integer. Prove that $\ker(T^k)$ reduces T if and only if $\ker(T^{k+1}) = \ker(T^k)$.
Hint: Both $\ker(T^k)$ and $\mathrm{range}(T^k)$ are T-invariant.

ex7.1.2 Let $T \in \mathscr{L}(\mathscr{V})$, and $\mathscr{V} = \mathscr{U} \oplus \mathscr{W}$ with both summands T-invariant. Let π be the projection onto $\mathscr{U}$ along $\mathscr{W}$. Prove that π commutes with T.

ex7.1.3 Prove that if $(\mathscr{V}, T)$ is irreducible, then its minimal polynomial is "prime power", that is, $\mathrm{minP_T} = \Phi^m$ with Φ irreducible and $m \geq 1$.

ex7.1.4 If $\mathscr{V}_j = \ker(\Phi_j^{m_j}(T))$ is a primary component of $(\mathscr{V}, T)$, the minimal polynomial of $T_{\mathscr{V}_j}$ is $\Phi_j^{m_j}$.

ex7.1.5 Assume that T is nonderogatory, i.e., $\chi_T = \mathrm{minP_T}$.

a. Prove that if $g \in \mathbb{F}[x]$ divides $\mathrm{minP_T}$, then $\dim \ker(g(T)) = \deg g$.

b. Prove that if $f \in \mathbb{F}[x]$, then

$$(7.1.12) \qquad\qquad \dim \ker(f(T)) = \deg \gcd(\mathrm{minP_T}, f).$$

Hint: Write $f(T) = g(T)h(T)$ where $g = \gcd(f, \mathrm{minP_T})$.

7.2   Semisimple systems

7.2.1   DEFINITION: The system $(\mathscr{V}, T)$ is *semisimple* if every T-invariant subspace of $\mathscr{V}$ is reducing.

Self-adjoint operators on inner-product spaces give the standard examples of semisimple systems.

Theorem. *The system* $(\mathcal{V}, T)$ *is semisimple if and only if* minP_T *is square-free (that is, the multiplicities* m_j *of the factors in the canonical factorization* $\mathrm{minP}_T = \prod_j \Phi_j^{m_j}$ *are all 1).*

PROOF: Proposition 7.1.6 reduces the general case to that in which minP_T is Φ^m with Φ irreducible.

If $m > 1$, then $\Phi(T)$ is not invertible, and hence the invariant subspace $\ker(\Phi(T))$ is nontrivial nor is it all of $\mathcal{V}$. $\ker(\Phi(T)^2)$ is strictly bigger than $\ker(\Phi(T))$ and, by Corollary 7.1.4, $\ker(\Phi(T))$ is not $\Phi(T)$-reducing, and hence not T-reducing.

If $m = 1$, we observe first that $\mathrm{minP}_{T,v} = \Phi$ for every nonzero vector $v \in \mathcal{V}$. This is true since $\mathrm{minP}_{T,v}$ divides Φ and Φ is prime. It follows that the dimension of $\mathrm{span}[T, v]$ is equal to the degree d of Φ, and hence: *every nontrivial T-invariant subspace has dimension $\geq d$.*

Let $\mathcal{W} \subset \mathcal{V}$ be a proper T-invariant subspace, and $v_1 \in \mathcal{V} \setminus \mathcal{W}$. The subspace $\mathrm{span}[T, v_1] \cap \mathcal{W}$ is T-invariant and is properly contained in $\mathrm{span}[T, v_1]$ (since $v_1 \notin \mathcal{W}$), so its dimension is less than d. This means that $\mathrm{span}[T, v_1] \cap \mathcal{W} = \{0\}$, so $\mathrm{span}[T, \mathcal{W} \cup \{v_1\}] = \mathcal{W} \oplus \mathrm{span}[T, v_1]$.

Let $\mathcal{W}_1 = \mathcal{W} \oplus \mathrm{span}[T, v_1]$. $\mathcal{W}_1$ is T-invariant and if $\mathcal{W}_1 \neq \mathcal{V}$, then by the preceding argument, if $v_2 \in \mathcal{V} \setminus \mathcal{W}_1$, then

$$\mathrm{span}[T, \mathcal{W}_1 \cup \{v_2\}] = \mathcal{W} \oplus \mathrm{span}[T, v_1] \oplus \mathrm{span}[T, v_2].$$

After $k = (\dim \mathcal{V} - \dim \mathcal{W})/d$ repetitions of this process, we have

(7.2.1) $$\mathcal{V} = \mathcal{W} \oplus \bigoplus_{j=1}^{k} \mathrm{span}[T, v_j],$$

where $\bigoplus_{1}^{k} \mathrm{span}[T, v_j]$ is clearly T-invariant, so $\mathcal{W}$ is reducing. ◀

Remark: If we start with $\mathcal{W} = \{0\}$, the decomposition (7.2.1) of $(\mathcal{V}, T)$ is a direct sum of cyclic subsystems.

7.2.2 If $\mathbb{F}$ is algebraically closed, then the irreducible polynomials in $\mathbb{F}[x]$ are linear. In other words, the prime factors Φ_j of minP_T have the form $x - \lambda_j$, with $\lambda_j \in \sigma(T)$.

By Theorem 7.2.1, if $(\mathcal{V}, T)$ is *semisimple*, then

(7.2.2) $\mathrm{minP}_T(x) = \displaystyle\prod_{\lambda_j \in \sigma(T)} (x - \lambda_j)$

and the *canonical prime-power decomposition* has the form

(7.2.3) $\mathcal{V} = \bigoplus \ker(T - \lambda_j) = \bigoplus \varphi_j(T)\mathcal{V},$

where $\varphi_j(T)$ are the projections defined in (7.1.8). The restriction of T to the eigenspace $\ker(T - \lambda_j) = \varphi_j(T)\mathcal{V}$ is just multiplication by λ_j, so that

(7.2.4) $T = \sum \lambda_j \varphi_j(T),$

and for every polynomial P,

(7.2.5) $P(T) = \sum P(\lambda_j)\varphi_j(T).$

A union of bases of the respective eigenspaces $\ker(T - \lambda_j)$ is a basis for $\mathcal{V}$ whose elements are all eigenvectors, and the matrix of T relative to this basis is diagonal. This proves the following theorem.

Theorem. *A semisimple operator on a vector space over an algebraically closed field is diagonalizable.*

7.2.3   DEFINITION: An algebra $\mathcal{B} \subset \mathcal{L}(\mathcal{V})$ is *semisimple* if every $T \in \mathcal{B}$ is semisimple. $\mathcal{B}$ is *diagonalizable* if $\mathcal{V}$ has a basis whose elements are all common eigenvectors of all the operators in $\mathcal{B}$.

Equivalently, $\mathcal{B}$ is *diagonalizable* if $\mathcal{V}$ has a basis relative to which the matrices of all the elements of $\mathcal{B}$ are diagonal.

Theorem. *Assume that $\mathbb{F}$ is algebraically closed and let $\mathcal{B} \subset \mathcal{L}(\mathcal{V})$ be a commutative semisimple subalgebra. Then*

a. $\mathcal{B}$ *is diagonalizable.*

b. $\mathcal{B}$ *is linearly spanned by the projections it contains.*

c. $\mathcal{B}$ *is singly generated, that is: there are elements $T \in \mathcal{B}$ such that $\mathcal{B} = \mathcal{P}(T) = \{P(T) : P \in \mathbb{F}[x]\}$.*

PROOF:

a. Let $T \in \mathcal{B}$ be such that $\sigma(T)$ is maximal (in terms of the number of its elements) and let

$$(7.2.6) \qquad \mathcal{V} = \bigoplus_{\lambda \in \sigma(T)} \mathcal{V}_\lambda$$

be the canonical decomposition of $\mathcal{V}$, where $\mathcal{V}_\lambda = \ker(T - \lambda)$.

Each $\mathcal{V}_\lambda$ is invariant under every operator S in $\mathcal{B}$, and the restrictions $S_\lambda = S_{\mathcal{V}_\lambda}$ are semisimple. We claim that S_λ acts as a scalar on $\mathcal{V}_\lambda$.

To see this, first note that if $\mu \in \sigma(S_\lambda)$, then $\lambda + a\mu \in \sigma(T + aS)$, for any $a \in \mathbb{F}$. Suppose now that for some $S \in \mathcal{B}$ and some $\lambda \in \sigma(T)$, the spectrum of S_λ has more than one element. Since $\mathbb{F}$ is infinite,[3] it follows from exercise **ex7.2.1** that we can find an $a \in \mathbb{F}$ such that $\sigma(T + aS)$ has more elements than $\sigma(T)$. This contradicts the choice of T.

Thus, for all $S \in \mathcal{B}$ and $\lambda \in \sigma(T)$, S_λ acts as scalar, so every element of $\mathcal{V}_\lambda$ is an eigenvector for every operator in $\mathcal{B}$. Take a basis in each $\mathcal{V}_\lambda$; their union is the required basis of $\mathcal{V}$.

b. Let $T \in \mathcal{B}$ be an operator with maximal spectrum, as above, and let $\mathcal{V} = \oplus \mathcal{V}_\lambda$ be the corresponding canonical decomposition into the eigenspaces of T. By Theorem 7.1.5, the projection onto $\mathcal{V}_\lambda$ is a polynomial in T, $\varphi_\lambda(T)$, so these projections are all in $\mathcal{B}$.

By part **a**, for any $S \in \mathcal{B}$ and $\lambda \in \sigma(T)$, there is a scalar μ_λ such that $Sv = \mu_\lambda v$ for all $v \in \mathcal{V}_\lambda$. It follows that

$$(7.2.7) \qquad S = \sum_{\lambda \in \sigma(T)} \mu_\lambda \cdot \varphi_\lambda(T).$$

c. The right-hand side of (7.2.7) is a polynomial in T. ◀

Remarks: The projections $\{\varphi_\lambda(T) : \lambda \in \sigma(T)\}$ (for T with maximal spectrum) form a basis for $\mathcal{B}$, see exercise **ex7.2.4**.

---

[3]Recall that an algebraically closed field must be infinite.

Different approaches to proving parts **b** and **c** are given in exercises
ex7.2.5 and **ex7.2.6**, respectively.

EXERCISES FOR SECTION 7.2

ex7.2.1 Let $(\lambda_1, \mu_1), \ldots, (\lambda_k, \mu_k)$ be k *distinct* points in $\mathbb{F}^2$. Show that there
are only finitely many $a \in \mathbb{F}$ such that $\lambda_i + a\mu_i = \lambda_j + a\mu_j$, for some pair
$1 \le i < j \le k$.

ex7.2.2 If T is diagonalizable ($\mathscr{V}$ has a basis consisting of eigenvectors of
T) then $(\mathscr{V}, T)$ is semisimple.

ex7.2.3 Let $\mathscr{V} = \bigoplus \mathscr{V}_j$ be an arbitrary direct sum decomposition of $\mathscr{V}$, and
let π_j be the corresponding projections. Let $T = \sum j\pi_j$.

d. Prove that T is semisimple.

e. Exhibit polynomials P_j such that $\pi_j = P_j(T)$.

ex7.2.4 Show that the projections $\{\varphi_\lambda(T) : \lambda \in \sigma(T)\}$ that appear in the
proof of part **b** of Theorem 7.2.3 are linearly independent.
Hint: See the remark that follows the proof of Theorem 7.1.5.

ex7.2.5 Let $\mathscr{B} \subset \mathscr{L}(\mathscr{V})$ be a commutative subalgebra. For projections in $\mathscr{B}$,
write $\pi_1 \le \pi_2$ if $\pi_1\pi_2 = \pi_1$.

a. Prove that this defines a partial order on the set of projections in $\mathscr{B}$.
Hint: Check that:

    i. if $\pi_1 \le \pi_2$ and $\pi_2 \le \pi_1$ then $\pi_1 = \pi_2$;
    ii. if $\pi_1 \le \pi_2$ and $\pi_2 \le \pi_3$, then $\pi_1 \le \pi_3$.

b. A projection $\pi \neq 0$ is *minimal* if $0 \neq \pi_1 \le \pi$ implies $\pi_1 = \pi$. Prove that
every projection in $\mathscr{B}$ is the sum of minimal projections.

c. Prove that if $\mathscr{B}$ is semisimple, then the set of minimal projections is a
basis for $\mathscr{B}$.

ex7.2.6 Prove part **c** of Theorem 7.2.3, by using the *Lagrange interpolation
theorem*, see exercise **ex3.1.8**, part **c**.
Hint: Using the notation in the proof of the theorem, show that $S = P(T)$,
where $P \in \mathbb{F}[x]$ is the polynomial satisfying $P(\lambda) = \mu_\lambda$, for $\lambda \in \sigma(T)$.

ex7.2.7 Let $\mathscr{V} = \mathscr{V}_1 \oplus \mathscr{V}_0$ and let $\mathscr{B} \subset \mathscr{L}(\mathscr{V})$ be the set of all the operators
S such that $S\mathscr{V}_1 \subset \mathscr{V}_0$, and $S\mathscr{V}_0 = \{0\}$. Prove that $\mathscr{B}$ is a commutative subalgebra of $\mathscr{L}(\mathscr{V})$, and that $\dim \mathscr{B} = \dim \mathscr{V}_0 \dim \mathscr{V}_1$. When is $\mathscr{B}$ semisimple?

ex7.2.8 Let $\mathscr{B}$ be the subset of $\mathscr{M}(2;\mathbb{R})$ of the matrices of the form $\begin{bmatrix} a & b \\ -b & a \end{bmatrix}$.
Prove that $\mathscr{B}$ is an algebra over $\mathbb{R}$, and is in fact a field isomorphic to $\mathbb{C}$.

ex7.2.9 Let $\mathscr{V}$ be an n-dimensional real vector space, and $T \in \mathscr{L}(\mathscr{V})$ an operator such that $\mathrm{minP}_T(x) = Q(x) = x^2 + bx + c$, with $b^2 < 4c$.

Prove that n is even and, for an appropriate basis in $\mathscr{V}$, the matrix A_T consists of $n/2$ copies of $\begin{bmatrix} \alpha & \beta \\ -\beta & \alpha \end{bmatrix}$ along the diagonal, where $\alpha = b/2$ and $\beta = \sqrt{4c - b^2}$.

In what sense is $(\mathscr{V}, T)$ "isomorphic" to $(\mathbb{C}^{n/2}, \lambda I)$? ($I$ the identity on $\mathbb{C}^{n/2}$).

7.3   Nilpotent operators

The canonical prime-power decomposition (see subsection 7.1.6) reduces every linear system to a direct sum of systems whose minimal polynomials are each a power of an irreducible polynomial Φ.

If $\mathbb{F}$ is algebraically closed, and in particular if $\mathbb{F} = \mathbb{C}$, the irreducible polynomials are linear, equal to $(x - \lambda)$ for some scalar $\lambda \in \sigma(T)$.

We consider here the (very important) special case in which all the irreducible factors Φ are linear. In other words, we focus on irreducible systems whose minimal polynomial has the form $\mathrm{minP}_T = (x - \lambda)^m$.

We discuss the general case in section $*7.5$.

7.3.1   If $\mathrm{minP}_T = (x - \lambda)^m$, then $\mathrm{minP}_{(T-\lambda)} = x^m$. As T and $T - \lambda I$ have the same invariant subspaces and the structure of $T - \lambda I$ tells everything about that of T, we focus on the case $\lambda = 0$.

DEFINITION: An operator $T \in \mathscr{L}(\mathscr{V})$ is *nilpotent* if $T^k = 0$ for some positive integer k. The smallest positive k for which $T^k = 0$ is called the *height* of T and denoted $\mathrm{height}[T]$. We refer to a linear system $(\mathscr{V}, T)$ as nilpotent of height k if T is nilpotent and $\mathrm{height}[T] = k$.

If $T^k = 0$, minP_T divides x^k, hence it is a power of x.

If $\mathrm{height}[T] = k$, then $T^{k-1} \neq 0$ and $\mathrm{minP}_T(x) = x^k$. In other words, T is *nilpotent of height k* if and only if $\mathrm{minP}_T(x) = x^k$.

For every $v \in \mathscr{V}$, $\mathrm{minP}_{T,v}(x) = x^l$ for an appropriate l. The *height* of v (under the action of T), denoted $\mathrm{height}[v]$, is the degree of $\mathrm{minP}_{T,v}$, that is, the smallest integer l such that $T^l v = 0$; it is the height of $T_{\mathcal{W}}$, where $\mathcal{W} = \mathrm{span}[T, v]$. Clearly $\mathrm{height}[T] = \max_{v \in \mathcal{V}} \mathrm{height}[v]$.

Since for $v \neq 0$, $\mathrm{height}[Tv] = \mathrm{height}[v] - 1$, elements of maximal height are not in $\mathrm{range}(T)$.

7.3.2 DEFINITION. A *k-shift* is a cyclic nilpotent system $(\mathcal{V}, T)$ of height k. A *standard shift* is a k-shift for some k, that is, a cyclic nilpotent system.

If $\{\mathcal{V}, T\}$ is a k-shift, and $v_0 \in \mathcal{V}$ is a vector of height k, then $\mathbf{v} = \{T^j v_0\}_{j=0}^{k-1}$ is a basis for $\mathcal{V}$, and the action of T is to map each basis element, except for the last, to the next one, and map the last basis element to 0. The matrix of T with respect to this basis is

$$(7.3.1) \qquad A_{T,\mathbf{v}} = \begin{bmatrix} 0 & 0 & \cdots & 0 & 0 \\ 1 & 0 & \cdots & 0 & 0 \\ 0 & 1 & \cdots & 0 & 0 \\ \vdots & \vdots & & \vdots & \vdots \\ 0 & 0 & \cdots & 1 & 0 \end{bmatrix}.$$

In particular, the dimension of $\mathcal{V}$ is k, equal to the height.

EXAMPLE: $\mathcal{V}$ is the space of all (algebraic) polynomials of degree bounded by m (so that $\{x^j\}_{j=0}^m$ is a basis for $\mathcal{V}$) and T is the differentiation operator:

$$(7.3.2) \qquad T\left(\sum_0^m a_j x^j\right) = \sum_1^m j a_j x^{j-1} = \sum_0^{m-1} (j+1) a_{j+1} x^j.$$

The vector $w = x^m$ has height $m + 1$, and $\{T^j w\}_{j=0}^m$ is a basis for $\mathcal{V}$ (so that w is a cyclic vector).

If we want the matrix of T to have the form (7.3.1), we have to normalize the basis elements, say $v_j = \frac{x^{m-j}}{(m-j)!}$, so that $v_{j+1} = T v_j$.

7.3.3   Shifts are the building blocks of nilpotent systems.

Theorem (Cyclic decomposition of nilpotent operators). *Let $(\mathcal{V}, T)$ be a finite-dimensional nilpotent system of height k. Then*

$$(7.3.3) \qquad\qquad \mathcal{V} = \bigoplus \mathcal{V}_j,$$

where $\mathcal{V}_j$ are T-invariant, and each $(\mathcal{V}_j, T_{\mathcal{V}_j})$ is a standard shift.

Moreover, for every $h \in \mathbb{N}$, the number $n(h)$ of summands $\mathcal{V}_j$ of height h in the decomposition (7.3.3) is uniquely determined.

PROOF: We use induction on $k = \text{height}[(\mathcal{V}, T)]$.

a. If $k = 1$, then $T = 0$ and any decomposition $\mathcal{V} = \bigoplus \mathcal{V}_j$ into one-dimensional subspaces will do.

b. Assume that the statement is valid for systems of height less that k and let $(\mathcal{V}, T)$ be a finite-dimensional nilpotent system of height k.

Write $\mathcal{W}_{\text{in}} = \ker(T) \cap T\mathcal{V}$, and let $\mathcal{W}_{\text{out}} \subset \ker(T)$ be a complementary subspace, i.e., $\ker(T) = \mathcal{W}_{\text{in}} \oplus \mathcal{W}_{\text{out}}$.

The system $(T\mathcal{V}, T)$ is nilpotent of height $k - 1$ and, by the induction hypothesis, admits a decomposition

$$T\mathcal{V} = \bigoplus_{j=1}^{m} \tilde{\mathcal{V}}_j$$

into standard shifts. For every $j \in [1, m]$ denote $l_j = \text{height}[(\tilde{\mathcal{V}}_j, T)]$. Let $\tilde{v}_j \in \tilde{\mathcal{V}}_j$ be of height l_j, so that $\tilde{\mathcal{V}}_j = \text{span}[T, \tilde{v}_j]$, and observe that $\{T^{l_j-1}\tilde{v}_j\}$ is a basis for $\mathcal{W}_{\text{in}}$.

Let v_j be such that $\tilde{v}_j = T v_j$, and write

$$\mathcal{V}_j = \text{span}[T, v_j] = \text{span}[\tilde{\mathcal{V}}_j, v_j].$$

Let $\mathcal{W}_{\text{out}} = \bigoplus_{i \leq l} \mathcal{W}_i$ be a direct sum decomposition of $\mathcal{W}_{\text{out}}$ into one-dimensional subspaces. The claim now is:

(7.3.4) $$\mathcal{V} = \bigoplus \mathcal{V}_j \oplus \bigoplus \mathcal{W}_i.$$

Observe that (7.3.4) is the same as (7.3.3), with some summands written as $\mathcal{W}_i$.

In order to prove (7.3.4) we need to verify that the set of spaces $\{\mathcal{V}_j\}_{j=1}^{m} \cup \{\mathcal{W}_i\}_{i=1}^{l}$ is independent, and their union spans $\mathcal{V}$.

Independence: Assume there is a nontrivial relation $\sum u_j + \sum w_i = 0$ with $u_j \in \mathcal{V}_j$ and $w_i \in \mathcal{W}_i$. Let $h = \max \text{height}[u_j]$.

If $h > 1$, then $\sum T^{h-1} u_j = T^{h-1}\left(\sum u_j + \sum w_i\right) = 0$ and we obtain a nontrivial relation between the $\tilde{\mathcal{V}}_j$'s. A contradiction.

If $h = 1$, we obtain a nontrivial relation between elements of a basis of $\ker(T)$. Again a contradiction.

Spanning: Denote $\mathcal{U} = \mathrm{span}[\{\mathcal{W}_i, \mathcal{V}_j\}]$, $i = 1, \ldots, l$, $j = 1, \ldots, m$. Then $T\mathcal{U}$ contains every $\tilde{v}_j$, and hence $T\mathcal{U} = T\mathcal{V}$. It folows that $\mathcal{U} \supset \mathcal{W}_{\mathrm{in}}$, and since it contains (by its definition) $\mathcal{W}_{\mathrm{out}}$, we have $\mathcal{U} \supset \ker(T)$.

For arbitrary $v \in \mathcal{V}$, let $\hat{v} \in \mathcal{U}$ be such that $Tv = T\hat{v}$. Then $v - \hat{v} \in \ker(T) \subset \mathcal{U}$, so that $v \in \mathcal{U}$, and $\mathcal{U} = \mathcal{V}$.

Finally, to prove that $n(h)$, the number of summands $\mathcal{V}_j$ of height h in the decomposition, is independent of the way it is obtained, observe that if $(\mathcal{W}, T)$ is an h-shift, the dimension of $T^l \mathcal{W}$ is $h - l$ if $h \geq l + 1$, and is equal to 0 if $h \leq l$. It follows that the number of nontrivial summands remaining in $T^l \mathcal{V}$ is precisely $\dim T^l \mathcal{V} - \dim T^{l+1} \mathcal{V}$, and $n(h)$ is the number of summands for $T^{h-1} \mathcal{V}$ which disappear in $T^h \mathcal{V}$, that is,

$$(7.3.5) \qquad n(h) = \dim T^{h-1} \mathcal{V} - 2 \dim T^h \mathcal{V} + \dim T^{h+1} \mathcal{V}. \qquad \blacktriangleleft$$

7.3.4 DEFINITION: A *cyclic decomposition* of a system $(\mathcal{V}, T)$ is a direct sum decomposition of the system into *irreducible cyclic subspaces*, that is, irreducible subspaces of the form $\mathrm{span}[T, v]$.

Theorem 7.3.3 allows, for systems whose minimal polynomial have only linear prime factors, a refinement of the canonical prime-power decomposition to a cyclic decomposition.

If $\mathrm{minP_T} = \prod_{\lambda \in \sigma(T)} (x - \lambda)^{m_\lambda}$, then the canonical prime-power decomposition, (7.1.7), gives

$$(7.3.6) \qquad \mathcal{V} = \bigoplus_{\lambda \in \sigma(T)} \ker((T - \lambda)^{m_\lambda}).$$

We apply Theorem 7.3.3 to each $\ker((T - \lambda)^{m_\lambda})$, λ in $\sigma(T)$, denote by $H_\lambda = H_\lambda(T)$ the sequence of the heights obtained in the cyclic decomposition of $\ker((T - \lambda)^{m_\lambda})$, each height h repeated $n_\lambda(h)$ times, and obtain the following theorem.

Theorem. *If $(\mathscr{V},T)$ is a linear system whose minimal polynomial has only linear factors, then*

$$(7.3.7) \qquad \mathscr{V} = \bigoplus_{\lambda \in \sigma(T),\, h \in H_\lambda} \mathscr{V}_{\lambda,h}$$

where $(\mathscr{V}_{\lambda,h}, T)$ is a cyclic subspace of height h of $\ker((T-\lambda))$.

EXERCISES FOR SECTION 7.3

ex7.3.1 In the notation of subsection 7.3.3, prove that

$$(7.3.8) \qquad \dim T^l \mathscr{V} = \sum_{h=l+1}^{k} (h-l)n(h).$$

ex7.3.2 The sequences $\{\dim T^l \mathscr{V}\}$, and $\{n(h)\}$ are (each) a complete similarity invariant for T.

ex7.3.3 Assume $\mathbb{F} = \mathbb{R}$ and $\minP_T = \Phi(x) = x^2 + 1$. Prove that the map $a + bT \mapsto a + bi$ is a (field) isomorphism of $\mathbb{F}_\Phi$ onto $\mathbb{C}$.

ex7.3.4 A *unipotent matrix* is a matrix A such that $A - I$ is nilpotent. Equivalent condition is "the minimal polynomial of A is $(x-1)^m$ for some m" or, if the underlying field is algebraically closed, "the spectrum of A reduces to $\{1\}$".

a. Prove that an upper-triangular matrix whose diagonal entries are all 1 is unipotent.

b. Prove that if $A \in \mathscr{M}(n,\mathbb{C})$ is unipotent, then A^j has polynomial growth; more precisely: there exists a constant K such that all the entries of A^j are bounded in absolute value by $K j^{m-1}$, where $m = \text{height}[A-I]$.

7.4 The Jordan canonical form

7.4.1 Bases and corresponding matrices. We consider now the case of a cyclic linear system $(\mathscr{V},T)$ whose minimal polynomial has the form $\minP_T = (x-\lambda)^h$ with $\lambda \in \mathbb{F}$.

If v is a cyclic vector, i.e., $\mathscr{V} = \text{span}[T,v]$, then $\minP_{T,v} = (x-\lambda)^h$, and the set $\{T^j v\}_{j=0}^{h-1}$ is a basis for $\mathscr{V}$, so that $\mathscr{V}$ is h-dimensional.

The system $(\mathscr{V}, (T-\lambda))$ is an h-shift, that is, nilpotent of height h, and v, having height h, is a cyclic vector for the system $(\mathscr{V}, (T-\lambda))$ as well. The matrix of $(T - \lambda)$ in terms of the basis $\mathbf{v} = \{(T - \lambda)^j v\}_{j=0}^{h-1}$ has the form (7.3.1) and the matrix of $T = (T - \lambda) + \lambda$ is the $h \times h$ matrix

(7.4.1)
$$J_{\lambda,h} = \begin{bmatrix} \lambda & 0 & 0 & \cdots & 0 & 0 \\ 1 & \lambda & 0 & \cdots & 0 & 0 \\ 0 & 1 & \lambda & \cdots & 0 & 0 \\ \vdots & \vdots & \ddots & \ddots & \vdots & \vdots \\ 0 & 0 & \cdots & 1 & \lambda & 0 \\ 0 & 0 & \cdots & 0 & 1 & \lambda \end{bmatrix},$$

that has all its diagonal entries equal to λ, all the entries just below the diagonal (assuming $h > 1$) are equal to 1, and all other entries are 0.

If we use such bases in every summand of the cyclic decomposition, (7.3.7), and take as a basis for $\mathcal{V}$ the union of these bases in successive blocks, each ordered as above, we obtain a basis of $\mathcal{V}$ with respect to which the matrix of T is the diagonal sum of matrices of the form (7.4.1).

Theorem (Jordan canonical form). *Let $(\mathcal{V}, T)$ be a linear system such that the prime factors of minP_T are all linear. In particular, $(\mathcal{V}, T)$ can be an arbitrary linear system over an algebraically closed field.*

Then there is a basis $\mathbf{v}$ for $\mathcal{V}$ such that the matrix $A_{T,\mathbf{v}}$ of T with respect to $\mathbf{v}$ is a diagonal sum of matrices $J_{\lambda,h}$, $\lambda \in \sigma(T)$, $h \in H_\lambda$.

EXERCISES FOR SECTION 7.4

ex7.4.1 Prove that $M_\lambda = \sum_{h \in H_\lambda} h$ is the exponent of $(x - \lambda)$ in the charactristic polynomial $\chi_T(x)$.

∗7.5 The cyclic decomposition, general case

Recall that a *cyclic decomposition* of a system $(\mathcal{V}, T)$ is a direct sum decomposition of the system into *irreducible cyclic subspaces*, that is, irreducible subspaces of the form $\mathrm{span}[T, v]$.

We have seen that when the underlying field is algebraically closed, and, more generally, if all the prime factors of $\mathrm{minP_T}$ are linear, then the canonical prime-power decomposition can be refined to a *cyclic decomposition*. We now show that this can done for every linear system.

The summands in the canonical prime-power decomposition have the form $\ker(\Phi^m(T))$ with an irreducible polynomial Φ. We show here that such systems (whose minimal polynomial is Φ^m, with irreducible Φ) admit a cyclic decomposition.

In section 7.3 we proved the claim for nilpotent operators, that is, the special case in which $\Phi(x) = x$ (and which extends immediately to the general linear case $\Phi(x) = x - \lambda$ for some $\lambda \in \mathbb{F}$).

The proof given below repeats, essentially verbatim, the proof given for the nilpotent case. Observe that the nilpotent operator now is $\Phi(T)$.

7.5.1   We assume that $\mathrm{minP_T} = \Phi^m$ and Φ is irreducible of degree d. For every $v \in \mathcal{V}$, $\mathrm{minP_{T,v}} = \Phi^{k(v)}$, $1 \le k \le m$, and $\max_v k(v) = m$; we refer to $k(v)$ as the Φ-height, or simply height, of v, and to m as the Φ-height, or simply height, of the system.

Theorem. *There exist vectors $v_j \in \mathcal{V}$ such that $\mathcal{V} = \bigoplus \mathrm{span}[T, v_j]$. Moreover, the set of the Φ-heights of the v_j's is uniquely determined.*

PROOF: We use induction on the Φ-height m.

a. $m = 1$. The system is semisimple and the canonical decomposition for it is cyclic. See section 7.2.

b. Assume that $\mathrm{minP_T} = \Phi^m$, $m > 1$, and the statement of the theorem valid for heights lower than m.

Write $\mathcal{W}_{\mathrm{in}} = \ker(\Phi(T)) \cap \Phi(T)\mathcal{V}$. The system $(\ker(\Phi(T)), T)$ is semisimple, see section 7.2, so that $\mathcal{W}_{\mathrm{in}}$ reduces $\ker(\Phi(T))$: there exists a T-invariant subspace $\mathcal{W}_{\mathrm{out}} \subset \ker(\Phi(T))$ complementary to $\mathcal{W}_{\mathrm{in}}$ i.e., such that $\ker(\Phi(T)) = \mathcal{W}_{\mathrm{in}} \oplus \mathcal{W}_{\mathrm{out}}$.

$(\Phi(T)\mathcal{V}, T)$ is of Φ-height $m - 1$ and, by the induction hypothesis, admits a decomposition $\Phi(T)\mathcal{V} = \bigoplus_{j=1}^{m} \tilde{\mathcal{V}}_j$ into cyclic subspaces, $\tilde{\mathcal{V}}_j = \mathrm{span}[T, \tilde{v}_j]$. Let v_j be such that $\tilde{v}_j = \Phi(T)v_j$.

Write $\mathscr{V}_j = \mathrm{span}[T, v_j]$, and let $\mathscr{W}_{\mathrm{out}} = \bigoplus_{i \leq l} \mathscr{W}_i$ be a direct sum decomposition into cyclic subspaces. The claim now is

(7.5.1) $$\mathscr{V} = \bigoplus \mathscr{V}_j \oplus \bigoplus \mathscr{W}_i.$$

To prove (7.5.1) we need to show that the spaces $\{\mathscr{V}_j, \mathscr{W}_i\}$, where $i = 1, \ldots, l$ and $j = 1, \ldots, m$, are independent, and that they span $\mathscr{V}$.

Independence: Assume that there is a nontrivial relation

$$\sum u_j + \sum w_i = 0 \text{ with } u_j \in \mathscr{V}_j \text{ and } w_i \in \mathscr{W}_i.$$

Let $h = \max \Phi\text{-height}[u_j]$.

If $h > 1$, then $\sum \Phi(T)^{h-1} u_j = \Phi(T)^{h-1} \left(\sum u_j + \sum w_i \right) = 0$ and we obtain a nontrivial relation between the $\widetilde{\mathscr{V}}_j$'s. A contradiction.

If $h = 1$, we obtain a nontrivial relation between elements of a basis of $\ker(\Phi)(T)$. Again a contradiction.

Spanning: Denote $\mathscr{U} = \mathrm{span}[\{\mathscr{W}_i, \mathscr{V}_j\}]$, $i = 1, \ldots, l$ and $j = 1, \ldots, m$. Notice first that $\mathscr{U} \supset \ker(T)$.

$\Phi(T) \mathscr{U}$ contains every $\tilde{v}_j$, and hence $T\mathscr{U} = T\mathscr{V}$. For $v \in \mathscr{V}$, let $\tilde{v} \in \mathscr{U}$ be such that $Tv = T\tilde{v}$. Then $v - \tilde{v} \in \ker(T) \subset \mathscr{U}$ so that $v \in \mathscr{U}$, and $\mathscr{U} = \mathscr{V}$.

Finally, just as in the previous subsection, we denote by $n(h)$ the number of v_j's of Φ-height h in the decomposition. Then $dn(m) = \dim \Phi(T)^{m-1}\mathscr{V}$ and, for $l = 0, \ldots, m-2$, we have, as in (7.3.5),

(7.5.2) $dn(h) = \dim \Phi(T)^{h-1}\mathscr{V} - 2\dim \Phi(T)^h \mathscr{V} + \dim \Phi(T)^{h+1}\mathscr{V}.$

◀

7.5.2 The cyclic decomposition.

We now refine the canonical prime-power decomposition (7.1.7):

$$\mathscr{V} = \bigoplus_{j=1}^{k} \ker(\Phi_j^{m_j}(T)),$$

by applying Theorem 7.5.1 to each of the summands $\ker(\Phi_j^{m_j}(T))$ and obtain the following theorem.

Theorem (General cyclic decomposition). *Let* $(\mathcal{V}, T)$ *be a linear system over a field* $\mathbb{F}$. *Let* $\min P_T = \prod_j \Phi_j^{m_j}$ *be the prime-power decomposition of its minimal polynomial. Then* $(\mathcal{V}, T)$ *admits a cyclic decomposition*

$$\mathcal{V} = \bigoplus \mathcal{V}_k.$$

To each k corresponds an index $j = j(k)$ such that $\mathcal{V}_k$ is a direct summand of $\ker(\Phi_j^{m_j})$, *so that the minimal polynomial of* $T_{\mathcal{V}_k}$ *is equal to* $\Phi_{j(k)}^{l(k)}$ *for some* $l(k) \le m_{j(k)}$, *and* $m_{j(k)} = \max l(k)$.

The polynomials $\Phi_{j(k)}^{l(k)}$ are called the *elementary divisors* of T.

Remark: We defined a *cyclic decomposition* as one in which the summands are *irreducible*. The requirement of irreducibility is satisfied automatically if the minimal polynomial is a "prime-power", i.e., has the form Φ^m with irreducible Φ. If one omits this requirement and the minimal polynomial has several relatively prime factors, we no longer have uniqueness of the decomposition since the direct sum of cyclic subspaces with relatively prime minimal polynomials is itself cyclic.

EXERCISES FOR SECTION 7.5

ex7.5.1 Assume $\min P_{T,v} = \Phi^m$ with irreducible Φ. Let $u \in \text{span}[T, v]$, and assume Φ-height$[u] = m$. Prove that $\text{span}[T, u] = \text{span}[T, v]$.

ex7.5.2 Give an example of two operators, T and S in $\mathcal{L}(\mathbb{C}^5)$, such that $\min P_T = \min P_S$ and $\chi_T = \chi_S$, and yet S and T are not similar.

ex7.5.3 Given 3 distinct irreducible polynomials Φ_j in $\mathbb{F}[x]$, $j = 1, 2, 3$. Let $\chi = \Phi_1^7 \Phi_2^3 \Phi_3^5$, $\Psi(x) = \Phi_1^3 \Phi_2^3 \Phi_3^3$, and denote

$$\mathcal{S}(\chi, \Psi) = \{T : T \in \mathcal{L}(\mathcal{V}), \ \min P_T = \Psi \ \text{and} \ \chi_T = \chi\}.$$

Assume that $\{T_k\}_{k=1}^N \subset \mathcal{S}(\chi, \Psi)$ is such that every element in $\mathcal{S}(\chi, \Psi)$ is similar to precisely one T_k. What is N?

ex7.5.4 Assume that $\mathbb{F}$ is a subfield of $\mathbb{F}_1$. Let $B_1, B_2 \in \mathcal{M}(n, \mathbb{F})$ and assume that they are $\mathbb{F}_1$-similar, i.e., $B_2 = C^{-1} B_1 C$ for some invertible $C \in \mathcal{M}(n, \mathbb{F}_1)$. Prove that they are $\mathbb{F}$-similar.

ex7.5.5 The operarors $T, S \in \mathcal{L}(\mathcal{V})$ are similar if and only if they have the same elementary divisors.

$*$7.6   The Jordan canonical form, general case

7.6.1   Bases and corresponding matrices. Let $(\mathscr{V}, T)$ be cyclic, that is, $\mathscr{V} = \operatorname{span}[T, v]$, and $\operatorname{minP}_T = \operatorname{minP}_{T,v} = \Phi^m$, with Φ irreducible of degree d. The cyclic decomposition provides several natural bases:

i.   The (ordered) set $\{T^j v\}_{j=0}^{dm-1}$ is a basis; the matrix of T with respect to this basis is the companion matrix of Φ^m.

ii.   Another natural basis in this context is

$$(7.6.1) \qquad \{T^k v\}_{k=0}^{d-1} \cup \{\Phi(T) T^k v\}_{k=0}^{d-1} \cup \cdots \cup \{\Phi(T)^{m-1} T^k v\}_{k=0}^{d-1};$$

the matrix A_{Φ^m} of T relative to this ordered basis consists of m copies of the companion matrix of Φ arranged on the diagonal, with 1's in the unused positions in the sub-diagonal.

If A_Φ is the companion matrix of Φ, then the matrix A_{Φ^4}, e.g., is

$$(7.6.2) \qquad A_{\Phi^4} =
\begin{pmatrix}
A_\Phi & & & \\
1 & A_\Phi & & \\
 & 1 & A_\Phi & \\
 & & 1 & A_\Phi
\end{pmatrix}.$$

7.6.2   The canonical form for real vector spaces.   When $(\mathscr{V}, T)$ is defined over $\mathbb{R}$, the irreducible factors Φ of minP_T are either linear or quadratic, i.e., have the form

$$\Phi(x) = x - \lambda, \qquad \text{or} \qquad \Phi(x) = x^2 + 2bx + c \quad \text{with} \quad b^2 - c < 0.$$

The companion matrix in the quadratic case is

$$(7.6.3) \qquad \begin{bmatrix} 0 & -c \\ 1 & -2b \end{bmatrix}.$$

(Over $\mathbb{C}$ we have $x^2 + 2bx + c = (x - \lambda)(x - \overline{\lambda})$ with $\lambda = -b + \sqrt{b^2 - c}$, and the matrix is similar to the diagonal matrix with λ and $\overline{\lambda}$ on the diagonal.)

EXERCISES FOR SECTION 7.6

ex7.6.1 Prove that every square complex-valued matrix is the product DU, where D is diagonal and U unipotent.

ex7.6.2 Assume that $v_1, \ldots, v_k$ are eigenvectors of T with the associated eigenvalues $\lambda_1, \ldots, \lambda_k$ all distinct. Prove that $v_1, \ldots, v_k$ are linearly independent.

ex7.6.3 Show that if we allow complex coefficients, the matrix (7.6.3) is similar to $\begin{bmatrix} \lambda & 0 \\ 0 & \overline{\lambda} \end{bmatrix}$ with $\lambda = -b + \sqrt{b^2 - c}$.

ex7.6.4 Assume that T is given by the matrix $A_T = \begin{bmatrix} 0 & 0 & 2 \\ 1 & 0 & 0 \\ 0 & 1 & 0 \end{bmatrix}$ acting on $\mathbb{F}^3$.

a. What is the basic decomposition when $\mathbb{F} = \mathbb{C}$, when $\mathbb{F} = \mathbb{R}$, and when $\mathbb{F} = \mathbb{Q}$?

b. Prove that when $\mathbb{F} = \mathbb{Q}$, every nonzero vector is cyclic. Hence, every nonzero rational vector is cyclic when $\mathbb{F} = \mathbb{R}$ or $\mathbb{C}$.

c. What happens to the basic decomposition under the action of an operator S that commutes with T?

d. Describe the set of matrices $A \in \mathcal{M}(3; \mathbb{F})$ that commute with A_T, where $\mathbb{F} = \mathbb{C}$, $\mathbb{R}$, and $\mathbb{Q}$ respectively.

ex7.6.5 Prove that the matrix $\begin{bmatrix} 0 & -1 \\ 1 & 0 \end{bmatrix}$ is not similar to a triangular matrix if the underlying field is $\mathbb{R}$, and is diagonalizable over $\mathbb{C}$. Why doesn't this contradict exercise **ex7.5.4**?

ex7.6.6 If $b^2 - c < 0$, then the (real) matrices $\begin{bmatrix} -b & \sqrt{c - b^2} \\ -\sqrt{c - b^2} & -b \end{bmatrix}$ and $\begin{bmatrix} 0 & -c \\ 1 & -2b \end{bmatrix}$ are similar.

ex7.6.7 Let $A \in \mathcal{M}(n; \mathbb{C})$ such that $\{A^j : j \in \mathbb{N}\}$ is bounded (under any norm on $\mathcal{M}(n; \mathbb{C})$; in particular: all the entries are uniformly bounded). Prove that all the eigenvalues of A are of absolute value not greater than 1. Moreover, if $\lambda \in \sigma(A)$ and $|\lambda| = 1$, there are no ones under λ in the Jordan canonical form of A.

ex7.6.8 Let $A \in \mathcal{M}(n;\mathbb{C})$ such that $\{A^j : j \in \mathbb{Z}\}$ is bounded. Prove that A is diagonalizable, and all its eigenvalues have absolute value 1.

ex7.6.9 Show that, with $A \in \mathcal{M}(n;\mathbb{C})$, the condition that $\{A^j : j \in \mathbb{N}\}$ is bounded is *not* sufficient to guarantee that A is diagonalizable.

However, if for some constant C and all polynomials $P \in \mathbb{C}[z]$, we have $\|P(A)\| \leq C \sup_{|z| \leq 1} |P(z)|$, then A is diagonalizable and all its eigenvalues have absolute values ≤ 1.

ex7.6.10 Let $T \in \mathcal{L}(\mathcal{V})$. Write $\chi_T = \prod \Phi_j^{m_j}$ where Φ_j are irreducible, but not necessarily distinct, and m_j are the corresponding heights in the cyclic decomposition of the system.

Find a basis of the form (7.6.1) for each of the components and describe the matrix of T relative to this basis.

ex7.6.11 Let $A = J_{\lambda,h}$ be the $h \times h$ matrix defined in (7.4.1). Compute A^n for all $n > 1$.

Hint: Write $A = \lambda I + B$.

Chapter 8

Additional Topics

Unless stated explicitly otherwise, the underlying field $\mathbb{F}$ of the vector spaces discussed in this chapter is either $\mathbb{R}$ or $\mathbb{C}$.

8.1 Functions of an operator

We assume in this section that the underlying field is $\mathbb{C}$.

If $P = \sum a_j x^j$ is a polynomial with coefficients in $\mathbb{F}$, we defined $P(T)$ by

$$P(T) = \sum a_j T^j.$$

The map $P \mapsto P(T)$ is a homomorphism of $\mathbb{F}[x]$ onto a subalgebra of $\mathscr{L}(\mathscr{V})$. We can often extend the homomorphism to a larger function space, but in most cases the range stays the same. The advantage will be in having a better match with the natural notation arising in applications.

8.1.1 Spectral approach. Write $\operatorname{minP_T}(z) = \prod_{\lambda \in \sigma(T)} (z - \lambda)^{m(\lambda)}$ and observe that a necessary and sufficient condition for a polynomial Q to be divisible by $\operatorname{minP_T}$ is that Q be divisible by $(z - \lambda)^{m(\lambda)}$ for every $\lambda \in \sigma(T)$, that is, have a zero of order at least $m(\lambda)$ at λ. It follows that $P_1(T) = P_2(T)$ if and only if the Taylor expansion of the two polynomials are the same up to, and including, the term of order $m(\lambda) - 1$ at every $\lambda \in \sigma(T)$.

In particular, if $m(\lambda) = 1$ for all $\lambda \in \sigma(T)$ (i.e., if $(\mathscr{V}, T)$ is semisimple), the condition $P_1(\lambda) = P_2(\lambda)$ for all $\lambda \in \sigma(T)$ is equivalent to $P_1(T) = P_2(T)$.

If F is an arbitrary numerical function defined on $\sigma(T)$ and T is semisimple, then the only consistent way to define $F(T)$ is by setting

$F(T) = P(T)$, where P is any polynomial that takes the same values as F at each point of $\sigma(T)$. This defines a homomorphism of the space of all numerical functions on $\sigma(T)$ onto the (same old) subalgebra generated by T in $\mathscr{L}(\mathscr{V})$.

In the general case, F needs to be defined and differentiable at least $m(\lambda) - 1$ times in a neighborhood of every $\lambda \in \sigma(T)$. With this assumption, we define $F(T) = P(T)$, where P is a polynomial whose Taylor expansion is the same as that of F up to, and including, the term of order $m(\lambda) - 1$ at every $\lambda \in \sigma(T)$.

8.1.2   Analytic approach.   The discussion in the previous subsection can only be put to use in practice if one has the complete spectral information about T. One needs to have the zeros of its minimal polynomial, including their multiplicities, given explicitly.

One can often define $F(T)$ without explicit knowledge of this information if F holomorphic in a sufficiently large set, and always if F is an entire function, that is, a function that admits a power series representation in the entire complex plane. This is done formally just as it was for polynomials, namely, for $F(z) = \sum_0^\infty a_n z^n$, we write $F(T) = \sum_0^\infty a_n T^n$, which is well defined (see *1.5 and *2.6) if the series converges. Since $\mathscr{L}(\mathscr{V})$ is finite-dimensional, so that all the norms on it are equivalent, we can use a submultiplicative "operator norm" as defined by (2.6.1). This keeps the estimates a little cleaner since $\|T^n\| \leq \|T\|^n$, and if the radius of convergence of the series is larger than $\|T\|$, the convergence of $\sum_0^\infty a_n T^n$ is assured.

EXAMPLES:

a.   Assume that the norm used is submultiplicative, and $\|T\| < 1$; then $(I - T)$ is invertible and $(I - T)^{-1} = \sum_{n=0}^\infty T^n$.

b.   Define $e^{zT} = \sum \frac{z^n}{n!} T^n$. The series is clearly convergent for every $T \in \mathscr{L}(\mathscr{V})$ and $a \in \mathbb{C}$. As a function of the parameter z it has the usual properties of the exponential function.

One may be tempted to ask whether e^{zT} has the same property as a function of T, that is, if $e^{z(T+S)} = e^{zT} e^{zS}$.

The answer is *yes* if S and T commute, but *no* in general; see exercise **ex8.1.4**.

8.1.3 In the context of the previous subsection, let $S_N(T) = \sum_0^N a_n T^n$. Division with remainder of S_N by $\min P_T$ gives

$$(8.1.1) \qquad\qquad S_N = q_N \min P_T + P_N,$$

with $\deg P_N < m = \deg \min P_T$, and we have $S_N(T) = P_N(T)$.

As $\|P_N(T) - P_M(T)\| = \|S_N(T) - S_M(T)\|$, the sequence $\{P_N(T)\}$ is a Cauchy sequence, and hence it converges to a limit $P(T) \in \mathscr{L}(\mathscr{V})$. As $\mathscr{P}(T)$ is a closed subspace of (the finite-dimensional) $\mathscr{L}(\mathscr{V})$, we have $P(T) \in \mathscr{P}(T)$, i.e., $P(T)$ is a polynomial in T of degree bounded by $m-1$, and

$$(8.1.2) \qquad F(T) = \lim_{N\to\infty} S_N(T) = \lim_{N\to\infty} P_N(T) = P(T).$$

The proof of the following proposition is left as an exercise.

Proposition. *Let F and T be as above and let P be the polynomial defined by (8.1.2). Then for all $\lambda \in \sigma(T)$*

$$(8.1.3) \qquad P^{(k)}(\lambda) = F^{(k)}(\lambda), \quad for\ 0 \le k \le m(\lambda) - 1.$$

The proposition shows that P is the same polynomial that we obtain using the spectral approach. Thus, the analytic approach can be thought of as a *practical* method for computing the value of $F(T)$ for analytic F when the spectral information of T is incomplete.

EXERCISES FOR SECTION 8.1

Assume that $\mathscr{V}$ is a finite-dimensional complex vector space.

ex8.1.1 Prove Proposition 8.1.3.

ex8.1.2 An operator $T \in \mathscr{L}(\mathscr{V})$ *has a square root* if there is $S \in \mathscr{L}(\mathscr{V})$ such that $T = S^2$.

a. Prove that every semisimple operator on $\mathbb{C}^n$ has a square root.

b. Prove that every invertible operator on a finite-dimensional complex vector space has a square root.

c.   Prove that the standard shift on $\mathbb{C}^n$ does not have a square root.

d.   Let T be the standard shift on $\mathbb{C}^3$. Find a square root for $I + T$.

e.   How many (distinct) semisimple square roots are there for the identity operator on an n-dimensional space $\mathscr{V}$? Can you find an operator $T \in \mathscr{L}(\mathscr{V})$ that has more square roots than the identity?

ex8.1.3   For a nonsingular $T \in \mathscr{L}(\mathscr{V})$ extend the definition of T^a from $a \in \mathbb{Z}$ to $a \in \mathbb{R}$ in a way that guarantees that for $a, b \in \mathbb{R}$, $T^{a+b} = T^a T^b$ (i.e., guarantees that $\{T^a\}_{a \in \mathbb{R}}$ is a one parameter subgroup).

ex8.1.4   Assume $T, S \in \mathscr{L}(\mathscr{V})$. Prove that

a. $e^{aT} e^{bT} = e^{(a+b)T}$.

b.   Prove that if S and T commute, then $e^{(T+S)} = e^T e^S$.

c.   Verify that $e^{(T+S)} \neq e^T e^S$ for $S = \begin{bmatrix} 0 & 0 \\ 1 & 0 \end{bmatrix}$ and $T = \begin{bmatrix} 0 & 1 \\ 0 & 0 \end{bmatrix}$.

ex8.1.5   Assume that $A \in \mathscr{M}(n; \mathbb{C})$. Show that each column $v(t)$ of e^{tA}, thought of as a vector of differentiable functions, is a solution of the linear system of differential equations $v'(t) = Av(t)$.

ex8.1.6   Let T denote the standard shift on $\mathbb{C}^n$. Find $\log(I + T)$.

ex8.1.7   Denote $\|T\|_\infty = \max_{\lambda \in \sigma(T)} |\lambda|$ (the *spectral norm of T*). Prove

(8.1.4) $$\|T\|_\infty \leq \liminf_{n \to \infty} \|T^n\|^{\frac{1}{n}}.$$

Hint: If $|\lambda| > \|T^k\|^{\frac{1}{k}}$ for some $k \in \mathbb{N}$, then the series $\sum_0^\infty \lambda^{-n} T^n$ converges.

Remark: The liminf appearing in (8.1.4) is in fact a limit. To see this, notice that $a_n = \log \|T^n\|$ is subadditive: $a_{n+m} \leq a_n + a_m$. This implies $a_{kn} \leq k a_n$, or $\frac{1}{kn} a_{kn} \leq \frac{1}{n} a_n$, for all $k \in \mathbb{N}$. This, in turn, implies $\lim \frac{1}{n} a_n = \liminf \frac{1}{n} a_n$.

8.2   Quadratic forms

8.2.1   A *quadratic form* in n variables is a polynomial $Q \in \mathbb{F}[x_1, \dots, x_n]$ of the form

(8.2.1) $$Q(x_1, \dots, x_n) = \sum_{i,j} a_{i,j} x_i x_j.$$

Since $x_i x_j = x_j x_i$, there is no loss of generality in assuming $a_{i,j} = a_{j,i}$.

A *Hermitian form* on an n-dimensional inner-product space $\mathscr{H}$ is a function of the form $Q(v) = \langle Tv, v \rangle$ with $T \in \mathscr{L}(\mathscr{H})$.

A basis $\mathbf{v} = \{v_1, \ldots, v_n\}$ transforms Q into a function $Q_{\mathbf{v}}$ of n variables on the underlying field, $\mathbb{R}$ or $\mathbb{C}$ as the case may be. We use the notation appropriate[1] for $\mathbb{C}$.

Write $v = \sum_1^n x_j v_j$ and $a_{i,j} = \langle Tv_i, v_j \rangle$; then $\langle Tv, v \rangle = \sum_{i,j} a_{i,j} x_i \bar{x}_j$ and

$$(8.2.2) \qquad Q_{\mathbf{v}}(x_1, \ldots, x_n) = \sum_{i,j} a_{i,j} x_i \bar{x}_j$$

expresses Q in terms of the variables $\{x_j\}$ (i.e., the $\mathbf{v}$-coordinates of v).

Conversely, given a function of the form (8.2.2), denote the matrix of coefficients $(a_{i,j})$ by $A_{\mathbf{v}}$, write $\mathbf{x} = \begin{bmatrix} x_1 \\ \vdots \\ x_n \end{bmatrix}$, and observe that

$$(8.2.3) \qquad Q_{\mathbf{v}}(x_1, \ldots, x_n) = \langle A\mathbf{x}, \mathbf{x} \rangle = \bar{\mathbf{x}}^{Tr} A_{\mathbf{v}} \mathbf{x}.$$

8.2.2 If we replace the basis $\mathbf{v}$ by another, say $\mathbf{w}$, the coefficients undergo a linear change of variables: there exists a matrix $C \in \mathscr{M}(n)$ that transforms by left-multiplication the $\mathbf{w}$-coordinates $\mathbf{y} = \begin{bmatrix} y_1 \\ \vdots \\ y_n \end{bmatrix}$ of a vector into its $\mathbf{v}$-coordinates: $\mathbf{x} = C\mathbf{y}$. Now

$$(8.2.4) \qquad Q_{\mathbf{v}}(x_1, \ldots, x_n) = \bar{\mathbf{x}}^{Tr} A_{\mathbf{v}} \mathbf{x} = \bar{\mathbf{y}}^{Tr} \overline{C}^{Tr} A_{\mathbf{v}} C \, \mathbf{y}$$

and the matrix representing Q in terms of the variables y_j, is[2]

$$(8.2.5) \qquad A_{\mathbf{w}} = \overline{C}^{Tr} A_{\mathbf{v}} C = C^* A_{\mathbf{v}} C.$$

DEFINITION: The matrices A and B are *congruent* if there exists a non-singular matrix C such that $B = C^* A C$.

Notice that the form now is $C^* A C$, rather than $C^{-1} A C$ (which defines similarity). The two notions agree if C is unitary, since then $C^* = C^{-1}$.

---

[1] If the underlying field is $\mathbb{R}$, the complex conjugation can simply be ignored.
[2] The adjoint of a matrix is introduced in 6.2.3.

8.2.3   Real-valued quadratic forms.   When the underlying field is $\mathbb{R}$, the quadratic form Q is real-valued. It does not determine the entries $a_{i,j}$ uniquely. Since $x_j x_i = x_i x_j$, the value of Q depends on $a_{i,j} + a_{j,i}$ and not on each of the summands separately. We may therefore assume, without modifying Q, that $a_{i,j} = a_{j,i}$, thereby making the matrix $A_\mathbf{v} = (a_{i,j})$ symmetric.

For real-valued quadratic forms over $\mathbb{C}$ the following lemma guarantees that the matrix of coefficients is Hermitian.

Lemma. *A quadratic form $\overline{\mathbf{x}}^{Tr} A_\mathbf{v} \mathbf{x}$ on $\mathbb{C}^n$ is real-valued if and only if the matrix of coefficients $A_\mathbf{v}$ is Hermitian,[3] i.e., $a_{i,j} = \overline{a_{j,i}}$.*

PROOF: If $a_{i,j} = \overline{a_{j,i}}$ for all i, j, then $\sum_{i,j} a_{i,j} x_i \overline{x}_j$ is it own complex conjugate.

Conversely, if we assume that $\sum_{i,j} a_{i,j} x_i \overline{x}_j \in \mathbb{R}$ for all $x_1, \ldots, x_n \in \mathbb{C}$, then:

Taking $x_j = 0$ for $j \neq k$, and $x_k = 1$, we obtain $a_{k,k} \in \mathbb{R}$. Taking $x_k = x_l = 1$ and $x_j = 0$ for $j \neq k, l$, we obtain $a_{k,l} + a_{l,k} \in \mathbb{R}$, that is, $\mathfrak{I}a_{k,l} = -\mathfrak{I}a_{l,k}$; while for $x_k = i, x_l = 1$ we obtain $i(a_{k,l} - a_{l,k}) \in \mathbb{R}$, i.e., $\mathfrak{R}a_{k,l} = \mathfrak{R}a_{l,k}$. Combining the two we have $a_{k,l} = \overline{a_{l,k}}$.   ◀

8.2.4   The fact that the matrix of coefficients of a real-valued quadratic form Q is self-adjoint makes it possible to simplify Q by a (unitary) change of variables that reduces it to a linear combination of squares. If the given matrix is A, we invoke the spectral theorem, Theorem 6.5.3, to obtain a unitary matrix U such that $U^* A U = U^{-1} A U$ is a diagonal matrix whose diagonal consists of the complete collection, including multiplicity, of the eigenvalues $\{\lambda_j\}$ of A. In other words, if $\mathbf{x} = U\mathbf{y}$, then

(8.2.6) $$Q(x_1, \ldots, x_n) = \sum \lambda_j |y_j|^2.$$

There are other matrices C which diagonalize Q, and the coefficients in the diagonal representation $Q(y_1, \ldots, y_n) = \sum b_j |y_j|^2$ depend on the one used. What does not depend on the particular choice of C

---

[3]Equivalently, if the operator T is self-adjoint.

is the number n_+ of positive coefficients, the number n_0 of zeros and the number n_- of negative coefficients. This is known as *The law of inertia.*

DEFINITION: A quadratic form $Q(v)$ on a (real or complex) vector space $\mathscr{V}$ is *positive definite* if $Q(v) > 0$ for all $v \neq 0$ in $\mathscr{V}$; it is *negative definite* if $Q(v) < 0$ for all $v \neq 0$ in $\mathscr{V}$.

If $\mathscr{V}$ is an inner-product space and $Q(v) = \langle Av, v \rangle$ with a self-adjoint operator A, our current definition is consistent with the definition in 6.6.1: the operator A is positive if $Q(v) = \langle Av, v \rangle$ is positive definite. We use the term *positive definite* to avoid confusion with *positive matrices* as defined in the following section.

Denote by n_+ the maximal dimension of subspaces $\mathscr{V}_1$ of $\mathscr{V}$ on which Q is positive definite, by n_- the maximal dimension of subspaces $\mathscr{V}_1$ of $\mathscr{V}$ on which Q is negative definite.

Proposition. *Let* **v** *be a basis in terms of which*

$$Q(y_1, \ldots, y_n) = \sum b_j |y_j|^2,$$

and arrange the coordinates so that $b_j > 0$ *for* $j \leq m$ *and* $b_j \leq 0$ *for* $j > m$. *Then* $m = n_+$.

PROOF: Denote $\mathscr{V}_+ = \text{span}[v_1, \ldots v_m]$, and $\mathscr{V}_{\leq 0} = \text{span}[v_{m+1}, \ldots v_n]$, the complementary subspace.

$Q(y_1, \ldots, y_n)$ is clearly positive on $\mathscr{V}_+$, so that $m \leq n_+$. On the other hand, by Theorem 2.5.3, every subspace $\mathscr{W}$ of dimension $> m$ has elements $v \in \mathscr{V}_{\leq 0}$, and for such v we clearly have $Q(v) \leq 0$.   ◀

The proposition applied to $-Q$ shows that n_- equals the number of negative b_j's. This proves

Theorem (Law of inertia). *Let Q be a real-valued quadratic form. Then in any representation $Q(y_1, \ldots, y_n) = \sum b_j |y_j|^2$, the number of positive coefficients is n_+, the number of negative coefficients is n_-, and the number of zeros is $n_0 = n - n_+ - n_-$.*

EXERCISES FOR SECTION 8.2

ex8.2.1 Prove that if $\langle Av, v \rangle = \langle Bv, v \rangle$ for all $v \in \mathbb{R}^n$, with $A, B \in \mathcal{M}(n, \mathbb{R})$, and both symmetric, then $A = B$.

ex8.2.2 Let $\{v_j\} \subset \mathscr{H}$. Write $a_{i,j} = \langle v_i, v_j \rangle$ and $A = (a_{i,j})$. Prove that A is positive definite if and only if $\{v_j\}$ is linearly independent.

ex8.2.3 The Gram determinant $|\Gamma|$ of the vectors v_j, $j = 1, \ldots, m$ in $\mathscr{H}$ is the detrminant of the matrix

$$\Gamma = \Gamma(v_1, \ldots, v_m) = \begin{bmatrix} \langle v_1, v_1 \rangle & \langle v_1, v_2 \rangle & \cdots & \langle v_1, v_m \rangle \\ \langle v_2, v_1 \rangle & \langle v_2, v_2 \rangle & \cdots & \langle v_2, v_m \rangle \\ \cdots & \cdots & \cdots & \cdots \\ \langle v_m, v_1 \rangle & \langle v_m, v_2 \rangle & \cdots & \langle v_m, v_m \rangle \end{bmatrix}.$$

a. Identify the inequality $\Gamma(v_1, v_2) \geq 0$.

b. Prove that $|\Gamma| \geq 0$, and it vanishes if and only if the vectors v_j are linearly dependent.

Hint: Identify the quadratic form $\langle \Gamma \mathbf{x}, \mathbf{x} \rangle$.

8.3   Perron-Frobenius theory

Matrices with positive, or more generally, nonnegative coefficients have wide application in a variety of fields. The basic spectral properties for such matrices are described by the theorems of Perron and Frobenius.

8.3.1   Notation and terminology.

DEFINITION: A matrix $A \in \mathcal{M}(n, \mathbb{C})$ is *positive* if all its entries are positive.[4] A is *nonnegative* if all its entries are nonnegative.

Similarly, a vector $v \in \mathbb{C}^n$ is *positive* if all its entries are positive, and *nonnegative*, if all its entries are nonnegative.

With A_j denoting either matrices or vectors, $A_1 \geq A_2, A_1 \gneq A_2$, and $A_1 > A_2$ will mean respectively that $A_1 - A_2$ is nonnegative, nonnegative but not zero, and positive.

Observe that if $A > 0$ and $v \gneq 0$, then $Av > 0$.

---

[4]Not to be confused with positivity, as defined in 6.6.1, of *the operator T_A* of multiplication by A.

The *spectral norm* of a matrix $A \in \mathcal{M}(m, \mathbb{C})$ is defined by

$$\|A\|_{sp} = \max\{|\tau| : \tau \in \sigma(A)\}.$$

In this section, it will be useful (and simplify notation) to think of $\mathbb{C}^n$ as the algebra of functions on the finite set $[1, \ldots, n]$, and we write vectors as $v = (v(1), \ldots, v(n))$. The *support* of a nonnegative vector v is the set of indices $j \in [1, \ldots, n]$ with $v(j) > 0$. For $v \in \mathbb{C}^n$, we denote by $|v|$ the vector $(|v(1)|, \ldots, |v(n)|)$. If $|v| > 0$ we denote by $\arg v$ and $e^{i \arg v}$ the vectors $(\arg v(1), \ldots, \arg v(n))$ and $(e^{i \arg v(1)}, \ldots, e^{i \arg v(n)})$, respectively. Also, for $u, v \in \mathbb{C}^n$ we denote the vector $(u(1)v(1), \ldots, u(n)v(n))$ by uv, (this should not be confused with the inner product).

8.3.2 Positive matrices. For a positive matrix A, we denote by $p(A)$ the set of all positive numbers μ for which there exist vectors $v \gneq 0$ such that

(8.3.1) $$Av \geq \mu v.$$

It is not hard to see that, on the one hand, $\min_i a_{i,i} \in p(A)$, and on the other, every $\mu \in p(A)$ is bounded by $\sum_{i,j} a_{i,j}$. Hence $p(A)$ is nonempty and bounded. Let $\rho = \sup_{\mu \in p(A)} \mu$.

Lemma. ρ *is an eigenvalue of A and has a positive corresponding eigenvector.*

PROOF: Let $\mu_n \in p(A)$ be such that $\mu_n \to \rho$, and let $v_n \gneq 0$ be such that $Av_n \geq \mu_n v_n$. We write, $v_n = (v_n(1), \ldots, v_n(m))$, and normalize v_n by the condition $\sum_j v_n(j) = 1$. Since now $0 \leq v_n(j) \leq 1$ for all n and j, we can choose a sequence n_k such that $v_{n_k}(j)$ converges for each $1 \leq j \leq m$. Denote the limits by $v_*(j)$ and let $v_* = (v_*(1), \ldots, v_*(m))$. We have $\sum_j v_*(j) = 1$, and since all the entries of Av_{n_k} converge to the corresponding entries in Av_*, we also have

(8.3.2) $$Av_* \geq \rho v_*,$$

so $\rho \in p(A)$.

If one of the entries in ρv_*, say $\rho v_*(l)$, were smaller than the l'th entry in Av_*, we could replace v_* by $v_{**} = v_* + \varepsilon \mathbf{e}_l$ (where $\mathbf{e}_l$ is the unit vector that has 1 as its l'th entry and zero everywhere else) with $\varepsilon > 0$ small enough to have

$$Av_*(l) \geq \rho v_{**}(l).$$

Since Ae_l is (strictly) positive, we would have $Av_{**} > Av_* \geq \rho v_{**}$, and for $\delta > 0$ sufficiently small we would have

$$Av_{**} \geq (\rho + \delta)v_{**}$$

contradicting the definition of ρ.

This shows that the inequality in (8.3.2) is in fact an equality, so ρ is an eigenvalue and v_* is a corresponding eigenvector. Since $v_* \gneq 0$ and $A > 0$, it follows that $\rho v_* = Av_* > 0$, so $v_* > 0$.   ◀

8.3.3 DEFINITION: An eigenvalue ρ of a matrix A is called *dominant* if

a. ρ is *simple*: $\ker((A - \rho)^2) = \ker(A - \rho)$ and $\dim \ker(A - \rho) = 1$; and

b. every other eigenvalue λ of A satisfies $|\lambda| < |\rho|$.
Notice that **b** implies that $|\rho| = \|A\|_{sp}$.

Theorem (Perron). *Let $A = (a_{i,j})$ be a positive matrix. Then it has a positive dominant eigenvalue and a positive corresponding eigenvector. Moreover, up to scalar multiplication, there is no other nonnegative eigenvector for A.*

PROOF: Let $\rho = \sup_{\mu \in p(A)} \mu$, then by Lemma 8.3.2, ρ is an eigenvalue of A with positive eigenvector v_*. Let $\lambda \neq \rho$ be another eigenvalue of A and w a corresponding eigenvector. The adjoint $A^* = \overline{A}^{tr}$ is a positive matrix and clearly, ρ is also an eigenvalue of A^* with a positive eigenvector, v^*. But $\langle w, v^* \rangle = 0$ (see exercise **ex6.2.4**), and since v^* is strictly positive, w cannot be nonnegative. Thus, to complete the proof of the theorem, it suffices to show that ρ is dominant.

First, we show that $\dim \ker(A - \rho) = 1$. If $Au = \rho u$ for some vector u, then the real and imaginary parts of u satisfy the same equality, so it suffices to show that if u has real entries then it is a constant multiple of v_*. Since $v_* > 0$, there exists a constant $c \neq 0$ such that $v_* + cu$ has all its entries nonnegative, and at least one vanishing entry. Now, $v_* + cu$ is an eigenvector for ρ and, unless $v_* + cu = 0$, we would have $\rho(v_* + cu) = A(v_* + cu) > 0$; this shows that $v_* + cu = 0$ and u is a multiple of v_*.

Next, we show that $\ker((A - \rho)^2) = \ker(A - \rho)$. Assume the contrary, and let $u \in \ker((A - \rho)^2) \setminus \ker(A - \rho)$, so that $(A - \rho)u$ is a nonzero element in $\ker(A - \rho)$. We have

$$(8.3.3) \qquad Au = \rho u + c v_*$$

with $c \neq 0$. Split (8.3.3) into its real and imaginary parts:

$$(8.3.4) \qquad A \Re u = \rho \Re u + \Re c v_* \quad A \Im u = \rho \Im u + \Im c v_*.$$

Either $c_1 = \Re c \neq 0$ or $c_2 = \Im c \neq 0$ (or both). This shows that there is no loss of generality in assuming that u and c in (8.3.3) are real-valued.

Replace u, if necessary, by $u_1 = -u$ to obtain $Au_1 = \rho u_1 + c_1 v_*$ with $c_1 > 0$. Since $v_* > 0$, we can choose $a > 0$ large enough to guarantee that $u_1 + a v_* > 0$, and observe that

$$A(u_1 + a v_*) = \rho(u_1 + a v_*) + c_1 v_*$$

so that $A(u_1 + a v_*) > \rho(u_1 + a v_*)$ contradicting the maximality of ρ.

Finally, let λ be an eigenvalue of A, and let $w \neq 0$ be a corresponding eigenvector: $Aw = \lambda w$. Denote $|w| = (|w(1)|, \ldots, |w(m)|)$.

The positivity of A implies $A|w| \geq |Aw|$ and

$$(8.3.5) \qquad A|w| \geq |Aw| \geq |\lambda||w|$$

so that $|\lambda| \in p(A)$, i.e., $|\lambda| \leq \rho$. If $|\lambda| = \rho$ we must have equality in (8.3.5) and $|w| = c v_*$. Equality in (8.3.5) can only happen if $A|w| = |Aw|$, which means that all the entries in w have the same argument, i.e., $w = e^{i\vartheta}|w|$. In other words, w is a constant multiple of v_*, so $\lambda = \rho$. ◀

8.3.4   Nonnegative matrices. Let III denote the matrix all of whose entries are 1. If $A \geq 0$ then $A + \frac{1}{m}\mathrm{III} > 0$ and has, by Perron's theorem, a dominant eigenvalue ρ_m and a corresponding positive eigenvector v_m, which we normalize by the condition $\sum_{j=1}^{n} v_m(j) = 1$.

ρ_m is monotone nonincreasing as $m \to \infty$ and converges to a limit $\rho = \|A\|_{sp} \geq 0$ (see $*$A.6.9). For a sequence $\{m_i\}$ the vectors v_{m_i} converge to a nonnegative vector v_* which, by continuity, is an eigenvector for ρ and satisfies $\sum v_*(l) = 1$.

That is, if A is a nonnegative matrix, then $\|A\|_{sp}$ is an eigenvalue of A with nonnegative eigenvector v_*. However, in general, there is no guarantee that $\|A\|_{sp}$ is dominant, nor indeed positive. Consider the following examples:

a. The identity matrix. 1 is the only eigenvalue and its multiplicity is equal to n.

b. The nilpotent matrix having ones below the diagonal, and zeros elsewhere. The spectrum is $\{0\}$.

c. The matrix A_σ of a permutation $\sigma \in S_n$. The spectrum depends on the decomposition of σ into cycles. If σ is a single cycle (of full length), then the spectrum of A_σ is the set of roots of unity of order n. The eigenvalue 1 has $(1, \ldots, 1)$ as a unique eigenvector. If the decomposition of σ consists of k cycles (including the trivial cycles) of lengths $l_j \geq 1$, $j = 1, \ldots, k$, then the spectrum of A_σ is the union of the sets of roots of unity of order l_j. The eigenvalue 1 now has multiplicity k.

Thus, for a general nonnegative matrix A

1. $\|A\|_{sp}$ may be zero;

2. $\|A\|_{sp}$ may have high multiplicity;

3. $\|A\|_{sp}$ may not have *positive* eigenvectors;

4. There may be other eigenvalues of modulus $\|A\|_{sp}$.

For a *transitive* nonnegative matrix A, the first three problems disappear, and the set of eigenvalues with modulus $\|A\|_{sp}$ has a simple algebraic structure. This is the subject of Frobenius' theorem, which follows below.

8.3.5 Definitions. Assume $A \geq 0$. We use the following terminology:

A connects the index j to i (connects (j,i)) *directly* if $a_{i,j} \neq 0$. Since $A\mathbf{e}_j = \sum a_{i,j}\mathbf{e}_i$, A connects (j,i) directly if $\mathbf{e}_i$ appears (with a nonzero coefficient) in the expansion of $A\mathbf{e}_j$.

More generally, *A connects* j to i (connects (j,i)) if, for some positive integer k, A^k connects j to i directly. This means: there is a *connecting chain for* (j,i), that is, a sequence $\{s_l\}_{l=0}^k$ such that $j = s_0$, $i = s_k$ and $\prod_{l=1}^k a_{s_l,s_{l-1}} \neq 0$. Notice that if a connecting chain for (j,i), $i \neq j$, has two occurrences of an index k, the part of the chain between the two is a *loop* that can be removed along with one k leaving a proper chain connecting (j,i). A chain with no loops has distinct entries and hence its length is $\leq n$. A chain which is itself a loop, that is, connecting an index to itself, can be similarly reduced to a chain of length $\leq n+1$. It follows from this that A connects (j,i) if and only if $B = \sum_{k=1}^n A^k$ connects (j,i) directly.

An index j is *A-recurrent* if A connects it to itself—there is a connecting chain for (j,j). The lengths k of connecting chains for (j,j) are called *return times* for j. Since connecting chains for (j,j) can be concatenated, the set of return times for a recurrent index is an additive semigroup of $\mathbb{N}$, (a subset that is closed under addition).

The existence of a recurrent index guarantees that $A^m \neq 0$ for all m; in other words—A is not nilpotent. This eliminates possibility **1** above.

8.3.6 Transitive matrices. The matrix A is *transitive* (also called *ergodic* or *irreducible*) if it connects every pair (j,i). If A is a nonnegative transitive matrix, every index is A-recurrent, A is not nilpotent, and $\rho = \|A\|_{sp} > 0$.

Since A connects (j,i) if and only if $B = \sum_{k=1}^n A^k$ connects (j,i) directly, it follows that a nonnegative matrix A is transitive if and only if B is positive. Since, by 8.3.4, ρ is an eigenvalue for A, it follows that $\beta = \sum_1^n \rho^j$ is an eigenvalue for B, having the same eigenvector v_*.

Applying Perron's theorem to B, it follows that v_* is in fact positive, every other nonnegative eigenvector of B is a multiple of v_* and β is the dominant eigenvalue for B. We note that $\beta = \|B\|_{sp}$ also follows directly from the spectral mapping theorem and the fact that $\rho = \|A\|_{sp}$.

Proposition. *If A is a transitive nonnegative matrix, then $\rho = \|A\|_{sp}$ is a simple eigenvalue of A and has a positive eigenvector v_*, which is the only nonnegative eigenvector of A, up to scalar multiplication.*

PROOF: Every eigenvector of A is an eigenvector of B, so any nonnegative eigenvector of A must be a multiple of v_*, and it follows that $\ker(A - \rho) = \ker(B - \beta)$ is 1-dimensional.

Next, observe that $B - \beta = \sum_{j=1}^n A^j - \rho^j = (A - \rho)Q(A)$, where Q is a polynomial of degree $n - 1$. This implies that

$$\ker((A - \rho)^2) \subseteq \ker((B - \beta)^2) = \ker((B - \beta)) = \ker((A - \rho)).$$

But $\ker((A - \rho)) \subseteq \ker((A - \rho)^2)$, so $\ker((A - \rho)^2) = \ker((A - \rho))$.
◀

8.3.7   Let A be a transitive nonnegative matrix, and let $\rho = \|A\|_{sp}$. By the proposition above, ρ is a simple eigenvalue of A, but it need not be dominant: A may have other eigenvalues of the same modulus.

Without loss of generality,[5] we may assume that $\rho = 1$, and normalize the corresponding positive eigenvector $v_* = (v_*(1), \dots, v_*(n))$ of A by the condition $\sum_j v_*(j) = 1$.

Lemma. *Assume that A is transitive, $v \geq 0$, $\mu > 0$, $Av \gneq \mu v$. Then there exists a positive vector $u \geq v$ such that $Au > \mu u$.*

PROOF: As in the proof of Perron's theorem: let l be an index such that $Av(l) > \mu v(l)$, let $0 < \varepsilon_1 < Av(l) - \mu v(l)$ and $v_1 = v + \varepsilon_1 \mathbf{e}_l$. Then $Av \geq \mu v_1$, hence

$$Av_1 = Av + \varepsilon_1 A\mathbf{e}_l \geq \mu v_1 + \varepsilon_1 A\mathbf{e}_l,$$

[5]By considering the transitive nonnegative matrix $\tilde{A} = \rho^{-1}A$.

and Av_1 is strictly bigger than μv_1 at l and at all the entries on which $A\mathbf{e}_l$ is positive, that is, the i's such that $a_{i,l} > 0$.

Now take $v_2 = v_1 + \varepsilon_2 A\mathbf{e}_l$ with $\varepsilon_2 > 0$ sufficiently small so that $Av_1 \geq \mu v_2$ and observe that

$$Av_2 = Av_1 + \varepsilon_2 A^2 \mathbf{e}_1 \geq \mu v_2 + \varepsilon_2 A^2 \mathbf{e}_l$$

and $Av_2 - \mu v_2$ is positive on all the entries on which $Av_1 - \mu v_1$ is positive, as well as on the support of $A^2 \mathbf{e}_l$; in particular on l and the support of $A\mathbf{e}_l + A^2 \mathbf{e}_l$. Continue in the same manner, taking $\varepsilon_3 > 0$ small enough, and $v_3 = v_2 + \varepsilon_3(A\mathbf{e}_l + A^2\mathbf{e}_l)$, so that $Av_2 \geq \mu v_3$ with strict inequality on the support of $(I + A + A^2 + A^3)\mathbf{e}_l$, etc. The transitivity of A guarantees that after $k \leq n$ such steps we obtain $u = v_k > 0$ such that $Au > \mu u$. ◀

The lemma implies in particular that if, for some $\mu > 0$, there exists a vector $v \geq 0$ such that $Av \gneq \mu v$, then $\mu < \rho$. This is true since the condition $Au > \mu u$ implies that $(A + \frac{1}{m}\mathrm{III})u > (1+a)\mu u$ for $a > 0$ sufficiently small,[6] and all m. This, in turn, implies that $\rho_m > (1+a)\mu$ for all m, and hence $\rho \geq (1+a)\mu$.

Proposition. *Assume* $\|A\|_{sp} = 1$. *If* $\xi = e^{i\varphi}$ *is an eigenvalue of* A *and* u_ξ *is a normalized eigenvector (that is,* $\sum_j |u_\xi(j)| = 1$*) corresponding to* ξ*, then*

a. $|u_\xi| = v_*$.

b. $|Au_\xi| = A|u_\xi|$.

PROOF: $$A|u_\xi| \geq |Au_\xi| = |\xi u_\xi| = |u_\xi|.$$

If $A|u_\xi| \neq |u_\xi|$ the lemma above would imply $\|A\|_{sp} > 1$, contradicting the assumption that $\|A\|_{sp} = 1$. Proposition 8.3.6 now implies part **a**, which in turn implies that both sides of part **b** are equal to v_*, and hence are equal. ◀

---

[6]See 8.3.4 for the notation.

8.3.8   Part **b** of Proposition 8.3.7 means that each entry in Au_ξ is a linear combination of entries of u_ξ *that have the same argument.* More precisley, we have the following structure.

The set $[1,\ldots,n]$ is partitioned into the level sets $I_{\vartheta(j)}$ such that for every $l \in I_{\vartheta(j)}$:

a. $\arg u_\xi(l) = \vartheta(j)$, and

b. A maps $\mathbf{e}_l$ into $\mathrm{span}[\{\mathbf{e}_k\}_{k \in I_{\vartheta(s)}}]$, where $\vartheta(s) = \vartheta(s) + \varphi$ ($\varphi = \arg\xi$).

In particular, A maps $\mathrm{span}[\{\mathbf{e}_l\}_{l \in I_{\vartheta(j)}}]$ into $\mathrm{span}[\{\mathbf{e}_k\}_{k \in I_{\vartheta(s)}}]$.

Let $\eta = e^{i\psi}$ be another eigenvalue of A, with corresponding eigenvector $u_\eta = e^{i \arg u_\eta} v_*$, and let $J_{\gamma(k)}$ be the level sets of $[1,\ldots,n]$ on which $\arg u_\eta = \gamma(k)$.

A maps every $\mathbf{e}_l$, for $l \in J_{\gamma(k)}$, into $\mathrm{span}[\{\mathbf{e}_m\}_{m \in J_{\gamma(t)}}]$ where $\gamma(t) = \gamma(k) + \psi$. It follows that if $l \in I_{\vartheta(j)} \cap J_{\gamma(k)}$, then

$$A\mathbf{e}_l \in \mathrm{span}[\{\mathbf{e}_k\}_{k \in I_{\vartheta(s)}}] \cap \mathrm{span}[\{\mathbf{e}_m\}_{m \in J_{\gamma(t)}}],$$

where $\vartheta(s) = \vartheta(j) + \varphi$ and $\gamma(t) = \gamma(k) + \psi$.   If we denote by $u_{\xi\eta}$ the vector $e^{i(\arg u_\xi + \arg u_\eta)} v_*$, then

$$\arg A e^{i(\vartheta(j)+\gamma(k))} \mathbf{e}_l = \arg u_\xi + \arg u_\eta + \varphi + \psi.$$

This being true for all $l \in [1,\ldots,n]$, it follows that $Au_{\xi\eta} = \xi\eta\, u_{\xi\eta}$.

Thus, if ξ and η are arbitrary eigenvalues of modulus 1 of A, then $\xi\eta = e^{i(\varphi+\psi)}$ is also an eigenvalue of A. This means that the set

$$\sigma(A)^* = \sigma(A) \cap \{z : |z| = 1\}$$

is a finite subgroup of the multiplicative group $\mathbb{T}^*$ of complex numbers of modulus 1. As such, $\sigma(A)^*$ is the group of roots of unity of order m, where m is equal to the cardinality of $\sigma(A)^*$.

For general transitive nonnegative A (i.e., when $\rho \neq 1$) we write $\sigma(A)^* = \{e^{it} : \rho e^{it} \in \sigma(A)\}$. In either case, $\sigma(A)^*$ is called the *period group* of A, and its order m is the *periodicity* of A.

If $\xi = e^{2\pi i/m}$ is the generator of the period group of A and u_ξ is an eigenvector corresponding to the eigenvalue $\rho e^{2\pi i/m}$, we call the partition of $[1, \ldots, n]$ into the level sets $I_{\vartheta(j)}$ of $\arg u_\xi$ the *basic partition*.

The subspaces $\mathscr{V}_j = \operatorname{span}[\{\mathbf{e}_l : l \in I_{\vartheta(j)}\}]$ are A^m-invariant and are mapped outside of themselves by A^k unless k is a multiple of m. The transitivity of A now implies that the restriction of A^m to $\mathscr{V}_j$ is transitive on $\mathscr{V}_j$, with the dominant eigenvalue 1, and $v_{*,j} = \sum_{l \in I_{\vartheta(j)}} v_*(l)\mathbf{e}_l$ is the corresponding eigenvector.

The restriction of A^m to $\mathscr{V}_j$ has (counting multiplicities) $|I_{\vartheta(j)}| - 1$ eigenvalues of modulus smaller than 1. Summing for $1 \leq j \leq m$ and invoking the spectral mapping theorem, Theorem 5.2.3, we see that A has $n - m$ eigenvalues of modulus < 1. This proves that the eigenvalues in the period group are simple and have no generalized eigenvectors.

Combining these observations with Proposition 8.3.6 yields the following theorem.

Theorem (Frobenius). *Let A be a transitive nonnegative $n \times n$ matrix. Then $\rho = \|A\|_{sp}$ is a simple eigenvalue of A and has a positive eigenvector v_*, which up to scalar multiplication is the only nonnegative eigenvector of A. Furthermore, the eigenvalues of modulus ρ are all simple, and the set $\sigma(A)^* = \{e^{it} : \rho e^{it} \in \sigma(A)\}$ is the group of roots of unity of order $m = |\sigma(A)^*|$.*

8.3.9 DEFINITION: A matrix $A \geq 0$ is *strongly transitive* if A^m is transitive for all $m \in [1, \ldots, n]$.

Theorem. *If A is strongly transitive, then $\|A\|_{sp}$ is a dominant eigenvalue for A, and has a positive corresponding eigenvector.*

PROOF: The periodicity of A has to be 1. ◄

8.3.10 The general nonnegative case. Let $A \in \mathscr{M}(n)$ be nonnegative. We write $i \leq_A j$ if A connects (i, j). This defines a partial order and induces an equivalence relation in the set of A-recurrent indices. (The nonrecurrent indices are not equivalent to themselves, nor to anything else.)

We can reorder the indices in a way that gives each equivalence class a consecutive bloc, and is compatible with the partial order, i.e., such that for nonequivalent indices, $i \leq_A j$ implies $i \leq j$. This ordering is not unique: equivalent indices can be ordered arbitrarily within their equivalence class; pairs of equivalence classes may be $\leq_A$ comparable or not comparable, in which case each may precede the other; nonrecurrent indices may be placed consistently in more than one place. Yet, such an order gives the matrix A a "quasi-super-triangular form": if we denote the coefficients of the "reorganized" A again by $a_{i,j}$, then $a_{i,j} = 0$ for i greater than the end of the bloc containing j. That means that now A has square transitive matrices centered on the diagonal— the squares $J_l \times J_l$ corresponding to the equivalence classes, while the entries on the rest of the diagonal, at the nonrecurrent indices, as well as in the rest of the subdiagonal, are all zeros.

This reduces much of the study of the general nonnegative A to that of transitive matrices.

EXERCISES FOR SECTION 8.3

ex8.3.1 What part of the conclusion of Perron's theorem remains valid if the assumption is replaced by "A is similar to a positive matrix" ?

ex8.3.2 Assume $A_1 \gneq A_2 > 0$, and let ρ_j be the dominant eigenvalues of A_j. Prove $\rho_1 > \rho_2$.

ex8.3.3 Let $A \in \mathcal{M}(n, \mathbb{R})$ be such that $P(A) > 0$ for some polynomial P in $\mathbb{R}[x]$. Prove that A has an eigenvalue $\lambda \in \mathbb{R}$ with positive eigenvector.
Hint: Use the spectral mapping theorem.

ex8.3.4 A nonnegative matrix A is nilpotent if and only if no index is A-recurrent.

ex8.3.5 Let A be a nonnegative matrix whose first row is positive, and let the $\leq_A$-equivalence class of $\mathbf{e}_1$ be $[1, \ldots, k]$. Show that A has a positive eigenvalue λ with a nonnegative eigenvector v whose entries $v(j)$ are positive for j in $[1, \ldots, k]$ and zero for $j > k$, yet there may be other positive eigenvalues larger than λ with corresponding nonnegative eigenvectors.
Hint: For the "yet" part consider $A = \begin{bmatrix} 1 & 2 \\ 0 & 2 \end{bmatrix}$.

ex8.3.6 Let A be a nonnegative matrix whose first row and first column are both positive. Prove that the properties guaranteed by Perron's theorem for positive matrices holds for A.

Would the same hold under the assumption that the first row and a column other than the first are both positive?

ex8.3.7 Prove that if the elements $I_{\vartheta(j)}$ of the basic partition are not equal in size, then $\ker(A)$ is nontrivial.

Hint: Show that $\dim\ker(A) \geq \max|I_{\vartheta(j)}| - \min|I_{\vartheta(j)}|$.

ex8.3.8 Describe the transitive matrix A if the basis elements are reordered so that the elements of the basic partition are blocs of consecutive integers in $[1,\ldots,n]$.

ex8.3.9 Prove that if $A \geq 0$ is transitive, then so is A^*.

ex8.3.10 Prove that if $A \geq 0$ is transitive, $\rho = \|A\|_{sp}$, and v^* is the positive eigenvector of A^*, normalized by the condition $\langle v_*, v^* \rangle = 1$, then for all $v \in \mathbb{C}^n$

$$(8.3.6) \qquad \lim_{N\to\infty} \frac{1}{N} \sum_1^N \rho^{-j} A^j v = \langle v, v^* \rangle v_*.$$

ex8.3.11 Let σ be a permutation of $[1,\ldots,n]$. Let A_σ be the $n \times n$ matrix whose entries a_{ij} are defined by

$$(8.3.7) \qquad a_{ij} = \begin{cases} 1 & \text{if } i = \sigma(j), \\ 0 & \text{otherwise.} \end{cases}$$

What is the spectrum of A_σ, and what are the corresponding eigenvectors?

ex8.3.12 Let $1 < k < n$, and let $\sigma \in S_n$ be the permutation consisting of the two cycles $(1,\ldots,k)$ and $(k+1,\ldots,n)$, and $A = A_\sigma$ as defined above. (So that the corresponding operator on $\mathbb{C}^n$ maps the basis vector e_i onto $e_{\sigma(i)}$.)

a. Describe the positive eigenvectors of A. What are the corresponding eigenvalues?

b. Let $0 < a, b < 1$. Denote by $A_{a,b}$ the matrix obtained from A by replacing the k'th and the n'th columns of A by $(c_{i,k})$ and $(c_{i,n})$, resp., where $c_{1,k} = 1 - a$, $c_{k+1,k} = a$ and all other entries (in the column) are zero; $c_{1,n} = b$, $c_{k+1,n} = 1 - b$ and all other entries are zero.

Show that 1 is a simple eigenvalue of $A_{a,b}$ and find a positive corresponding eigenvector. Show also that for other eigenvalues there are no nonnegative eigenvectors.

8.4   Stochastic matrices

8.4.1   A *stochastic matrix* is a nonnegative matrix $A = (a_{i,j})$ such that the sum of the entries in each column[7] is 1:

(8.4.1) $$\sum_i a_{i,j} = 1.$$

A *probability vector* is a nonnegative vector $\pi = (p(1), \ldots, p(n))$ such that $\sum_l p(l) = 1$. Observe that if A is a stochastic matrix and π a probability vector, then $A\pi$ is a probability vector.

In applications, one considers a set of possible outcomes of an "experiment" at a given time. The outcomes are often referred to as *states*, and a probability vector assigns probabilities to the various states. The word probability is taken here in a broad sense—if one is studying the distribution of various populations, the "probability" of a given population is simply its proportion in the total population.

A (stationary) *n-state Markov chain* is a sequence $\{v_j\}_{j \geq 0}$ of probability vectors in $\mathbb{R}^n$, such that

(8.4.2) $$v_j = Av_{j-1} = A^j v_0,$$

where A is an $n \times n$ stochastic matrix.

The matrix A is the *transition matrix*, and the vector v_0 is referred to as the *initial probability vector*. The parameter j is often referred to as *time*.

8.4.2   Positive transition matrix.   When the transition matrix A is positive, we get a clear description of the evolution of the Markov chain from Perron's theorem, Theorem 8.3.3.

Condition (8.4.1) is equivalent to $u^* A = u^*$, where u^* is the row vector $(1, \ldots, 1)$. This means that the dominant eigenvalue for A^* is 1, hence the dominant eigenvalue for A is 1. If v_* is the corresponding (positive) eigenvector, normalized so as to be a probability vector, then $Av_* = v_*$ and hence $A^j v_* = v_*$ for all j.

---

[7]The action of the matrix is (left) multiplication of column vectors. The columns of the matrix are the images of the standard basis in $\mathbb{R}^n$ or $\mathbb{C}^n$.

If w is another eigenvector (or generalized eigenvector), it is orthogonal to u^*, that is: $\sum_1^n w(j) = 0$. Also, $\sum |A^l w(j)|$ is exponentially small (as a function of l).

If v_0 is any probability vector, we write $v_0 = cv_* + w$ with w in the span of the (generalized) eigenspaces of the nondominant eigenvalues. By the remark above, $c = \sum v_0(j) = 1$. Then $A^l v_0 = v_* + A^l w$ and, since $A^l w \to 0$ as $l \to \infty$, we have $A^l v_0 \to v_*$.

Finding the vector v_* amounts to solving a homogeneous system of n equations (knowing a priori that the solution set is one-dimensional). The observation $v_* = \lim A^l v_0$, with v_0 an arbitrary probability vector, may be a fast way way to obtain a good approximation of v_*.

8.4.3 Transitive transition matrix. Denote by v_ξ the eigenvectors of A corresponding to eigenvalues ξ of absolute value 1, normalized so that $v_1 = v_*$ is a probability vector, and $|v_\xi| = v_*$. If the periodicity of A is m, then, for every probability vector v_0, the sequence $A^j v_0$ is equal to an m-periodic sequence (given by $A^j u_0$, u_0 being the component of v_0 in the span of the eigenvectors corresponding to eigenvalues of absolute value 1, all of which are m'th roots of unity) plus a sequence that tends to zero exponentially fast.

Since every eigenvalue $\xi \neq 1$ of absolute value 1 is an m'th root of unity, $\sum_1^m \xi^l = 0$. It follows that if v_0 is a probability vector, then

$$(8.4.3) \qquad \frac{1}{m} \sum_{l=k+1}^{k+m} A^l v_0 \to v_*$$

exponentially fast (as a function of k).

8.4.4 Reversible Markov chains. Given a nonnegative symmetric matrix $(p_{i,j})$, we write $W_j = \sum_i p_{ij}$ and, assuming $W_j > 0$ for all j, $a_{i,j} = \frac{p_{i,j}}{W_i}$. The matrix $A = (a_{i,j})$ is stochastic since $\sum_i a_{i,j} = 1$ for all j.

We can identify the "stable distribution"—the A-invariant vector— by thinking in terms of "population movement". Assume that at a given time we have population of size b_j in state j and in the next unit of time a proportion of size $a_{i,j}$ of this population shifts to state i. The absolute size of the population moving from j to i is $a_{i,j} b_j$ so that the new distribution is given by $A\mathbf{b}$, where $\mathbf{b}$ is the column vector with entries b_j. This description applies to any stochastic matrix, and the stable distribution is given by $\mathbf{b}$ which is invariant under A, $A\mathbf{b} = \mathbf{b}$.

The easiest way to find $\mathbf{b}$ in the present case is to go back to the matrix $(p_{i,j})$ and the weights W_j. The vector $\mathbf{w}$ with entries W_j is A-invariant in a very strong sense. Not only is $A\mathbf{w} = \mathbf{w}$, but the population exchange between any two states is even:

- the population moving from i to j is: $W_i a_{j,i} = p_{j,i}$;
- the population moving from j to i is: $W_j a_{i,j} = p_{i,j}$;
- the two are equal since $p_{i,j} = p_{j,i}$.

8.5   Representation of finite groups

Throughout this section G will denote a finite group.

8.5.1 DEFINITION. A *representation* of a group G in a vector space $\mathscr{V}$ is a *homomorphism*, $\sigma : g \mapsto T_g$, of G into the group $\mathbf{GL}(\mathscr{V})$ of invertible elements in $\mathscr{L}(\mathscr{V})$. The representation is *faithful* if σ is injective; it is finite-dimensional if the space $\mathscr{V}$ is finite-dimensional; the dimension of the representation is defined to be the dimension of $\mathscr{V}$.

A representation σ of G in $\mathscr{V}$ makes $\mathscr{V}$ into a *G-space* $(\mathscr{V}, G, \sigma)$, that is, a vector space for which, in addition to the vector space operations, there is an *action* of G on $\mathscr{V}$ by linear maps assigned by σ. This means that for every $g \in G$ there is an operator $\sigma(g) = T_g \in \mathscr{L}(\mathscr{V})$, and for all $g_1, g_2 \in G$ and $v \in \mathscr{V}$,

$$T_{g_1 g_2} v = T_{g_1} T_{g_2} v.$$

Note that the operators T_g are necessarily invertible, i.e., belong to $\mathbf{GL}(\mathscr{V})$, so that a G-space is simply a vector space with a given representation of G on it. We shall use the terms G-space and representation interchangeably.

If σ, or equivalently the action of G, is assumed known, we denote such spaces by $(\mathscr{V}, G)$ and often write $\mathbf{g}$ instead of $\sigma(g)$ or T_g.

If σ is a representation of G in $\mathscr{V}$, then a *G-subspace* of $\mathscr{V}$ is a vector subspace $\mathscr{W} \subset \mathscr{V}$ that is invariant under the action of G. This means that $\mathbf{g}w \in \mathscr{W}$ for all $g \in G$ and $w \in \mathscr{W}$. The restriction of (the operators assigned by) σ to $\mathscr{W}$ is called a *subrepresentation* of σ.

The following discussion is valid for abelian and nonabelian groups alike. The abelian case, however, is much simpler, and was given in 6.5.5 as an application of Theorem 6.3.6 (the spectral theorem for commutative, self-adjoint subalgebras).

8.5.2 The dual representation. If σ is a representation of G in $\mathscr{V}$, we obtain a representation σ^* of G in $\mathscr{V}^*$ by setting $\sigma^*(g) = \sigma(g^{-1})^*$ (the adjoint of the inverse of the action of G on $\mathscr{V}$). Since both $\mathbf{g} \mapsto \mathbf{g}^{-1}$ and $\mathbf{g} \mapsto \mathbf{g}^*$ reverse the order of factors in a product, their combination as used above preserves the order, and we have

$$\sigma^*(g_1 g_2) = \sigma^*(g_1)\sigma^*(g_2)$$

so that σ^* is in fact a homomorphism.

If the underlying field is $\mathbb{R}$ or $\mathbb{C}$, then we will assume that the space is equipped with an inner product. There is no loss of generality, since an inner product may always be introduced, e.g., by declaring a given basis to be orthonormal. We denote the G-space by $\mathscr{H}$ in these cases.

DEFINITION: A representation σ of G in a complex inner-product space $\mathscr{H}$ is *unitary* if $\sigma(g)$ is unitary for every $g \in G$.

If σ is a representation in an inner-product space $\mathscr{H}$, then the dual representation σ^* is also a representation in $\mathscr{H}$. If σ is *unitary*, then $\sigma^* = \sigma$.

8.5.3 Let $\mathscr{V}_1$ and $\mathscr{V}_2$ be G-spaces. We extend the actions of G on these spaces, denoted $\mathbf{g}_1$ and $\mathbf{g}_2$ respectively, to $\mathscr{V}_1 \oplus \mathscr{V}_2$ and $\mathscr{V}_1 \otimes \mathscr{V}_2$ by declaring

(8.5.1) $\mathbf{g}(v_1 \oplus v_2) = \mathbf{g}_1 v_1 \oplus \mathbf{g}_2 v_2$ and $\mathbf{g}(v_1 \otimes v_2) = \mathbf{g}_1 v_1 \otimes \mathbf{g}_2 v_2$.

We note that $\mathscr{L}(\mathscr{V}_1, \mathscr{V}_2) = \mathscr{V}_2 \otimes \mathscr{V}_1^*$ and as such it is a G-space.

8.5.4 *G-maps.* Let $\mathscr{V}_1$ and $\mathscr{V}_2$ be G-spaces. A map $S: \mathscr{V}_1 \to \mathscr{V}_2$ is *a G-map* if it is a linear map that commutes with the action of G. This means that for every $g \in G$, $S\mathbf{g}_1 = \mathbf{g}_2 S$, where $\mathbf{g}_i$ denotes the action of

G on $\mathscr{V}_i$. In other words, a linear map S is a G-map if and only if the diagram

$$\begin{array}{ccc} \mathscr{V}_1 & \xrightarrow{\ S\ } & \mathscr{V}_2 \\ \mathbf{g}_1 \downarrow & & \downarrow \mathbf{g}_2 \\ \mathscr{V}_1 & \xrightarrow{\ S\ } & \mathscr{V}_2 \end{array}$$

is commutative.

The prefix G- can be attached to all words describing linear maps; thus, a *G-isomorphism* is an isomorphism which is a G-map, etc.

If $\mathscr{V}_1$ and $\mathscr{V}_2$ are G-spaces, then we denote by $\mathscr{L}_G(\mathscr{V}_1, \mathscr{V}_2)$ the space of G-maps of $\mathscr{V}_1$ into $\mathscr{V}_2$.

DEFINITION: The representations $(\mathscr{V}_1, G)$ and $(\mathscr{V}_2, G)$ are *equivalent* if there is a G-isomorphism $S : \mathscr{V}_1 \mapsto \mathscr{V}_2$, that is, if they are isomorphic as G-spaces.

Proposition. *Let* $S : \mathscr{V}_1 \to \mathscr{V}_2$ *be a G-map. Then* $\ker(S)$ *is a subrepresentation of* $\mathscr{V}_1$, *and* $\mathrm{range}(S)$ *is a subrepresentation of* $\mathscr{V}_2$.

PROOF: If $v \in \ker(S)$, then $S\mathbf{g}_1 v = \mathbf{g}_2 S v = 0$, so $\mathbf{g}_1 v \in \ker(S)$. Likewise, if $w \in \mathrm{range}(S)$, then there is a $v \in \mathscr{V}_1$ such that $\mathbf{g}_2 w = \mathbf{g}_2 S v = S\mathbf{g}_1 v$, so $\mathbf{g}_2 w \in \mathrm{range}(S)$.   ◀

8.5.5   Averaging, I. Let $\mathscr{V}$ be a finite-dimensional space. For a finite subgroup $\mathscr{G} \subset \mathbf{GL}(\mathscr{V})$ we write

(8.5.2) $I_{\mathscr{G}} = \{v \in \mathscr{V} : \mathbf{g}v = v \text{ for all } \mathbf{g} \in \mathscr{G}\}$.

In words: $I_{\mathscr{G}}$ is the space of all the vectors in $\mathscr{V}$ which are invariant under every $\mathbf{g}$ in $\mathscr{G}$.

Theorem. *The operator*

(8.5.3) $$\pi_{\mathscr{G}} = \frac{1}{|\mathscr{G}|} \sum_{\mathbf{g} \in \mathscr{G}} \mathbf{g}$$

is a projection onto $I_{\mathscr{G}}$.

PROOF: $\pi_{\mathscr{G}}$ is clearly the identity on $I_{\mathscr{G}}$. All we need to do is show that $\text{range}(\pi_{\mathscr{G}}) = I_{\mathscr{G}}$, and for that observe that if $v = \frac{1}{|\mathscr{G}|} \sum_{\mathbf{g} \in \mathscr{G}} \mathbf{g}u$, then

$$\mathbf{g}_1 v = \frac{1}{|\mathscr{G}|} \sum_{\mathbf{g} \in \mathscr{G}} \mathbf{g}_1 \mathbf{g}u,$$

and since $\{\mathbf{g}_1 \mathbf{g} : \mathbf{g} \in \mathscr{G}\} = \mathscr{G}$, we have $\mathbf{g}_1 v = v$. ◀

8.5.6 Averaging, II. Let $\mathscr{H}$ be a finite-dimensional, complex inner-product space, and let $\mathscr{G}$ be a finite subgroup of $\mathbf{GL}(\mathscr{H})$.

The operator $Q = \frac{1}{|\mathscr{G}|} \sum_{\mathbf{g} \in \mathscr{G}} \mathbf{g}^* \mathbf{g}$ is self-adjoint and positive on $\mathscr{H}$, and can be used to define a new inner product:

$$(8.5.4) \qquad \langle v, u \rangle_Q = \langle Qv, u \rangle = \frac{1}{|\mathscr{G}|} \sum_{\mathbf{g} \in \mathscr{G}} \langle \mathbf{g}v, \mathbf{g}u \rangle,$$

with the corresponding norm

$$\|v\|_Q^2 = \frac{1}{|\mathscr{G}|} \sum_{\mathbf{g} \in \mathscr{G}} \langle \mathbf{g}v, \mathbf{g}v \rangle = \frac{1}{|\mathscr{G}|} \sum_{\mathbf{g} \in \mathscr{G}} \|\mathbf{g}v\|^2.$$

Since $\mathscr{G}\mathbf{h} = \{\mathbf{gh} : \mathbf{g} \in \mathscr{G}\} = \mathscr{G}$, for any $\mathbf{h} \in \mathscr{G}$, we have

$$(8.5.5) \qquad \langle \mathbf{h}v, \mathbf{h}u \rangle_Q = \frac{1}{|\mathscr{G}|} \sum_{\mathbf{g} \in \mathscr{G}} \langle \mathbf{gh}v, \mathbf{gh}u \rangle = \langle Qv, u \rangle,$$

and $\|\mathbf{h}v\|_Q = \|v\|_Q$. Thus, $\mathscr{G}$ is a subgroup of the unitary group corresponding to the new inner product $\langle \cdot, \cdot \rangle_Q$.

Denote by $\mathscr{H}_Q$ the inner-product space obtained by replacing the original inner product by $\langle \cdot, \cdot \rangle_Q$. Let $\{u_1, \ldots, u_n\}$ be an orthonormal basis of $\mathscr{H}$, and $\{v_1, \ldots, v_n\}$ an orthonormal basis of $\mathscr{H}_Q$. Define $S \in \mathbf{GL}(\mathscr{H})$ by setting $Su_j = v_j$ and extending by linearity.

So defined, S is an isometry from $\mathscr{H}$ onto $\mathscr{H}_Q$, with S^{-1} being the inverse isometry. Since $\mathbf{g}$ is unitary on $\mathscr{H}_Q$, for all $\mathbf{g} \in \mathscr{G}$, it follows that $S^{-1}\mathbf{g}S$ is unitary on $\mathscr{H}$. In other words, S conjugates $\mathscr{G}$ to a subgroup of the unitary group $U(\mathscr{H})$. This proves the following theorem

Theorem. *Every finite subgroup of* $\mathbf{GL}(\mathscr{H})$ *is conjugate to a sub-group of the unitary group* $U(\mathscr{H})$.

If $\sigma : G \to \mathbf{GL}(\mathcal{H})$ is a representation of a finite group in a complex space $\mathcal{H}$, then $\mathcal{G} = \sigma(G)$ is a finite subgroup of $\mathbf{GL}(\mathcal{H})$. Since we may assume that $\mathcal{H}$ is equipped with an inner product, the previous theorem implies the following corollary.

Corollary. *Every finite-dimensional representation of a finite group in a complex vector space is equivalent to a unitary representation.*

8.5.7   Let G be a finite group and $\mathcal{H}$ a finite-dimensional G-space.

DEFINITION: A subspace $\mathcal{U} \subset \mathcal{H}$ is *G-reducing*, or reducing for short, if it is G-invariant and has a G-invariant complement, that is, $\mathcal{H} = \mathcal{U} \oplus \mathcal{V}$, where both summands are G-invariant.

The representation $(\mathcal{H}, G)$ is *irreducible* if there is no nontrivial G-invariant subspace of $\mathcal{H}$ and *reducible* otherwise. In the terminology of **ex5.3.13**, $\mathcal{H}$ is irreducible if $(\mathcal{H}, \mathcal{G})$ is minimal.

Proposition. *If $\sigma : G \to \mathbf{GL}(\mathcal{H})$ is a representation of G in a finite-dimensional complex vector space, then every G-invariant subspace is reducing.*

PROOF: We assume, without loss of generality, that $\mathcal{H}$ is an inner-product space. Endow $\mathcal{H}$ with the inner product given by (8.5.4), with $\mathcal{G} = \sigma(G)$, making the representation unitary. Now, observe that if $\mathcal{U}$ is a nontrivial G-invariant subspace, then so is its orthogonal complement.   ◄

If $\mathcal{H} = \mathcal{U} \oplus \mathcal{V}$, and both $\mathcal{U}$ and $\mathcal{V}$ are G-invariant, then we say that the representation $(\mathcal{H}, G)$ is the *sum* of the representations $(\mathcal{U}, G)$ and $(\mathcal{V}, G)$.

Lemma. *Let $(\mathcal{V}, G)$ and $(\mathcal{U}, G)$ be irreducible subrepresentations of $(\mathcal{H}, G)$. Then, either $\mathcal{U} \cap \mathcal{V} = \{0\}$, or $\mathcal{U} = \mathcal{V}$.*

PROOF: $\mathcal{U} \cap \mathcal{V}$ is clearly G-invariant.   ◄

Theorem. *Every finite-dimensional complex representation $(\mathcal{H}, G)$ of a finite group G is a sum of irreducible representations. That is,*

$$(8.5.6) \qquad\qquad \mathcal{H} = \bigoplus \mathcal{U}_j,$$

where each $\mathcal{U}_j$ is an irreducible G-subspace of $\mathcal{H}$. Furthermore, the decomposition is unique.

PROOF: We prove the existence by induction on the dimension.

If $\dim \mathcal{H} = 1$, then there is nothing to prove. Assume now that the statement is valid for representations of dimension $< n$, and let $(\mathcal{H}, G)$ be a representation of dimension n.

If $(\mathcal{H}, G)$ is irreducible, then we are done. Otherwise, Proposition 8.5.7 shows that there is direct sum decomposition $\mathcal{H} = \mathcal{U} \oplus \mathcal{V}$ where both summands are nontrivial G-invariant subspaces. Since $\mathcal{U}$ and $\mathcal{V}$ have dimensions less than n, they each have a decomposition into irreducible G-subspaces, by the induction hypothesis. The sum of the two decompositions is a decomposition of $\mathcal{H}$.

The uniqueness follows from lemma 8.5.7. ◀

8.5.8 The regular representation. Let G be a finite group. Denote by $\ell^2(G)$ the vector space of all complex-valued functions on G, and define the inner product, for $\varphi, \psi \in \ell^2(G)$, by

$$\langle \varphi, \psi \rangle = \sum_{x \in G} \varphi(x) \overline{\psi(x)}.$$

For $g \in G$, the *left-translation by g* is the operator $\rho(g)$ on $\ell^2(G)$ defined by

$$(\rho(g)\varphi)(x) = \varphi(g^{-1}x).$$

Clearly $\rho(g)$ is linear and, in fact, unitary. Moreover,

$$(\rho(g_1 g_2)\varphi)(x) = \varphi((g_1 g_2)^{-1}x) = \varphi(g_2^{-1}(g_1^{-1}x)) = (\rho(g_1)\rho(g_2)\varphi)(x)$$

so that $\rho(g_1 g_2) = \rho(g_1)\rho(g_2)$ and ρ is a unitary representation of G. It is called the *regular representation of G*.

If $H \subset G$ is a subgroup, we denote by $\ell^2(G/H)$ the subspace of $\ell^2(G)$ of the functions that are constant on left cosets of H.

Since multiplication on the left by arbitrary $g \in G$ maps left H-cosets onto left H-cosets, $\ell^2(G/H)$ is $\rho(g)$-invariant. Unless G has *no* nontrivial subgroups—we say that G is *simple* in this case—ρ is reducible. This proves the following proposition.

Proposition. *If the regular representation of G is irreducible, then G is simple.*

The converse is false! A cyclic group of order p, with prime p, is simple. Yet, it follows from the discussion in subsection 6.5.5 that the regular representation of any finite abelian group is reducible to 1-dimensional summands.

8.5.9 Let $\mathcal{H}$ be a complex G-space and let $\langle\,,\rangle$ be an inner product in $\mathcal{H}$. Fix a nonzero vector $u \in \mathcal{H}$ and, for $v \in \mathcal{H}$ and $g \in G$, define

$$(8.5.7) \qquad\qquad f_v(g) = \langle \mathbf{g}^{-1}v, u\rangle.$$

The map $S\colon v \mapsto f_v$ is a linear map from $\mathcal{H}$ into $\ell^2(G)$.

Lemma. *If $\mathcal{H}$ is irreducible, then S is injective.*

PROOF: If $v \neq 0$, then the set $\{\mathbf{g}v : g \in G\}$ spans $\mathcal{H}$. This implies that f_v is not the 0 function, i.e., $S(v) \neq 0$. ◀

Observe that for $\gamma \in G$,

$$(8.5.8) \qquad \rho(\gamma)f_v(g) = f_v(\gamma^{-1}g) = \langle \mathbf{g}^{-1}\gamma v, u\rangle = f_{\gamma v}(g),$$

so that the space $S\mathcal{H} = \mathcal{H}_S \subset \ell^2(G)$ is a G-invariant subspace, and so, by Proposition 8.5.7, a reducing subspace of the regular representation of $\ell^2(G)$. Furthermore, S maps $(\mathcal{H}, G)$ onto $(\mathcal{H}_S, \rho|_{\mathcal{H}_S})$.

Together with the lemma above, this proves the following theorem.

Theorem. *Every irreducible representation of a finite group G in a finite-dimensional complex space is equivalent to a subrepresentation of the regular representation.*

Corollary. *There are only a finite number of distinct irreducible complex representations of a finite group G.*

Appendix

A.1 Equivalence relations–partitions

A.1.1 Binary relations. Formally, a *binary relation* R in a set X is a subset $R \subset X \times X$. It is the set of all pairs x, y such that x and y "have relation R". $(x, y) \in R$ is most often written as $x\overline{R}y$, where $\overline{R}$ is a symbol designating the relation.

EXAMPLES:

a. Equality: $R = \{(x,x) : x \in X\} = \{(x,y) : x, y \in X; x = y\}$.

b. Order in $\mathbb{Z}$: $R = \{(x,y) : x < y\}$.

c. Divisibility in $\mathbb{N}$: $R = \{(m,n) : m \mid n\}$, (m divides n in $\mathbb{N}$).

The symbol $\overline{R}$ in each of these is the usual one, i.e., $=$, $<$, and $\mid$ respectively.

A.1.2 Equivalence relations. An *equivalence relation* in a set X is a binary relation (denoted here $x \equiv y$) that is
reflexive: for all $x \in X$, $x \equiv x$;
symmetric: for all $x, y \in X$, if $x \equiv y$, then $y \equiv x$; and
transitive: for all $x, y, z \in X$, if $x \equiv y$ and $y \equiv z$, then $x \equiv z$.

EXAMPLES:

a. Of the binary relations above, *equality* is an equivalence relation; *order* and *divisibility* are not.

b. Congruence modulo an integer. Here $X = \mathbb{Z}$, the set of integers. Fix an integer k. We say that x *is congruent to y modulo k* and write $x \equiv y \pmod{k}$ if $x - y$ is an integer multiple of k.

c. For $X = \{(m,n) : m, n \in \mathbb{Z}, n \neq 0\}$, define $(m,n) \equiv (m_1, n_1)$ by the condition $mn_1 = m_1 n$. This will be familiar if we write the pairs as $\frac{m}{n}$ instead of (m,n) and observe that the condition $mn_1 = m_1 n$ is the one defining the equality of the rational fractions $\frac{m}{n}$ and $\frac{m_1}{n_1}$.

A.1.3   Partitions. A *partition of X* is a collection $\mathscr{P}$ of *pairwise disjoint* subsets $P_\alpha \subset X$ whose union is X, i.e.,

$$P_\alpha \cap P_\beta = \emptyset \ \text{ if } \ \alpha \neq \beta, \quad \text{and} \quad \bigcup_{\mathscr{P}} P_\alpha = X.$$

A partition $\mathscr{P}$ defines an equivalence relation: by definition, $x \equiv y$ if and only if x and y belong to the same element of the partition.

Conversely, given an equivalence relation on X, we define the *equivalence class* of $x \in X$ as the set $\mathscr{E}_x = \{y \in X : x \equiv y\}$. The defining properties of equivalence can be rephrased as:

a. $x \in \mathscr{E}_x$,
b. If $y \in \mathscr{E}_x$, then $x \in \mathscr{E}_y$, and
c. If $y \in \mathscr{E}_x$ and $z \in \mathscr{E}_y$, then $z \in \mathscr{E}_x$.

These conditions guarantee that different equivalence classes are disjoint and the collection of all the equivalence classes is a partition of X (that defines the given equivalence relation).

EXERCISES FOR SECTION A.1

exA.1.1 Let $R_1 \subset \mathbb{R} \times \mathbb{R} = \{(x,y) : |x-y| < 1\}$ and $x \sim_1 y$ when $(x,y) \in R_1$. Is this an equivalence relation, and if not—what fails?

exA.1.2 Congruence mod k is the relation $R = \{(m,n) \in \mathbb{Z}^2 : m - n \in k\mathbb{Z}\}$. Identify the equivalence classes for congruence mod k.

A.2   Maps

The terms used to describe properties of maps vary by author, by time, by subject matter, etc. We shall use the following:

A map $\varphi \colon X \to Y$ is *injective* if $x_1 \neq x_2 \implies \varphi(x_1) \neq \varphi(x_2)$. Equivalent terminology: φ is *one-to-one* (or 1-1), or φ is a *monomorphism*.

A map $\varphi \colon X \to Y$ is *surjective* if $\varphi(X) = \{\varphi(x) : x \in X\} = Y$. Equivalent terminology: φ is *onto*, or φ is an *epimorphism*.

A map $\varphi\colon X \to Y$ is *bijective* if it is both injective and surjective: for every $y \in Y$ there is precisely one $x \in X$ such that $y = \varphi(x)$. Bijective maps are *invertible*—the inverse map is defined by: $\varphi^{-1}(y) = x$ if $y = \varphi(x)$.

Maps that preserve some structure are called morphisms, often with a prefix providing additional information. Besides the *mono-* and *epi-* mentioned above, we use systematically *homomorphism, isomorphism*, etc.

A *permutation* of a set is a bijective map of the set onto itself.

A.3 Groups

A.3.1 DEFINITION: A *group* is a pair $(G, *)$, where G is a set and $*$ is a binary operation $(x, y) \mapsto x * y$, defined for all pairs $(x, y) \in G \times G$, taking values in G, and satisfying the following conditions:

G-1 The operation is associative: For $x, y, z \in G$, $(x * y) * z = x * (y * z)$.

G-2 There exists a unique element $e \in G$ called the *identity element* or the *unit* of G, such that $e * x = x * e = x$ for all $x \in G$.

G-3 For every $x \in G$ there exists a unique element x^{-1}, called *the inverse* of x, such that $x^{-1} * x = x * x^{-1} = e$.

A group $(G, *)$ is *abelian*, or *commutative*, if $x * y = y * x$ for all x and y. The group operation in a commutative group is often written and referred to as *addition*, in which case the identity element is written as 0, and the inverse of x as $-x$.

When the group operation is written as multiplication, the operation symbol $*$ is sometimes written as a dot (i.e., $x \cdot y$ rather than $x * y$) and is often omitted altogether. We also simplify the notation by referring to the group, when the binary operation is assumed known, as G rather than $(G, *)$.

The *order* of a group G, denoted $|G|$, is its cardinality.

EXAMPLES:

a. $(\mathbb{Z}, +)$, the integers with standard addition.

b. $(\mathbb{R} \setminus \{0\}, \cdot)$, the non-zero real numbers, standard multiplication.

c. S_n, the *symmetric group on* $[1, \ldots, n]$. Here n is a positive integer, the elements of S_n are all the permutations σ of the set $[1, \ldots, n]$, and the operation is *composition:* for $\sigma, \tau \in S_n$ and $1 \leq j \leq n$ we set $(\tau\sigma)(j) = \tau(\sigma(j))$.

More generally, if X is a set, the collection $S(X)$ of permutations, i.e., invertible self-maps of X, is a group under composition. (Thus $S_n = S([1, \ldots, n])$).

The first two examples are commutative; the third, if $n > 2$, is not.

A.3.2 Let G_i, $i = 1, 2$, be groups.

DEFINITION: A map $\varphi \colon G_1 \to G_2$ is a *homomorphism* if

(A.3.1) $$\varphi(xy) = \varphi(x)\varphi(y).$$

Notice that the multiplication on the left-hand side is in G_1, while that on the right-hand side is in G_2.

The definition of homomorphism is quite broad; we do not assume the mapping to be *injective* (1-1), nor *surjective* (onto). We use the proper adjectives explicitly whenever relevant: *monomorphism* for injective homomorphism and *epimorphism* for one that is surjective.

An *isomorphism* is a homomorphism which is *bijective*, that is, both injective and surjective. Bijective maps are invertible, and the inverse of an isomorphism is an isomorphism. For the proof we only have to show that φ^{-1} is multiplicative (as in (A.3.1)), that is, that for $g, h \in G_2$, $\varphi^{-1}(gh) = \varphi^{-1}(g)\varphi^{-1}(h)$. But, if $g = \varphi(x)$ and $h = \varphi(y)$, this is equivalent to $gh = \varphi(xy)$, which is the multiplicativity of φ.

If $\varphi \colon G_1 \to G_2$ and $\psi \colon G_2 \to G_3$ are both isomorphisms, then $\psi\varphi \colon G_1 \to G_3$ is an isomorphism as well.

We say that two groups G and G_1 are *isomorphic* if there is an isomorphism of one onto the other. The discussion above makes it clear that this is an equivalence relation.

A.3.3 Inner automorphisms and conjugacy classes. An isomorphism of a group onto itself is called an *automorphism*. A special class of automorphisms, the *inner automorphisms*, are the *conjugations by elements $y \in G$*:

(A.3.2) $$\varphi_y x = y^{-1} x y.$$

One checks easily (left as an exercise) that for all $y \in G$, the map φ_y is in fact an automorphism of G.

An important equivalence relation in G is *conjugacy*, defined by: $x \sim z$ if there exists $y \in G$ such that $z = \varphi_y x = y^{-1} x y$.

To check that every x is conjugate to itself take $y = e$, the identity. If $z = \varphi_y x$, then $x = \varphi_{y^{-1}} z$, proving the symmetry. Finally, if $z = y^{-1} x y$ and $u = w^{-1} z w$, then

$$u = w^{-1} z w = w^{-1} y^{-1} x y w = (yw)^{-1} x (yw),$$

which proves the transitivity.

The equivalence classes defined on G by conjugation are called *conjugacy classes*.

A.3.4 Subgroups and cosets.

DEFINITION: A subgroup of a group G is a subset $H \subset G$ such that

SG-1 H is closed under multiplication: if $h_1, h_2 \in H$ then $h_1 h_2 \in H$.

SG-2 If $h \in H$, then $h^{-1} \in H$.

Observe that these conditions imply that $e \in H$. In other words, $H \subset G$ is a subgroup if, with the operation inherited from G, it is a group.

EXAMPLES:

a. $\{e\}$, the subset whose only term is the identity element.

b. In $\mathbb{Z}$, the set $q\mathbb{Z}$ of all the integral multiples of some integer q. This is a special case of the following example.

c. For any $x \in G$, the set $\{x^k\}_{k \in \mathbb{Z}}$ is *the subgroup generated by x*. A group generated by one of its elements is called *cyclic*. The element *x* is *of order m* if the cyclic group it generates is of order *m*. (That is, if *m* is the smallest positive integer for which $x^m = e$.) *x* has infinite order if $\{x^n\}$ is infinite, in which case $n \mapsto x^n$ is an isomorphism of $\mathbb{Z}$ onto the group generated by *x*.

d. The subset of S_n of all the permutations that leave some (fixed) $l \in [1, \ldots, n]$ in its place, that is, $\{\sigma \in S_n : \sigma(l) = l\}$.

If $\varphi : G \to G_1$ is a homomorphism and e_1 denotes the identity in G_1, then $\{g \in G : \varphi g = e_1\}$ is a subgroup of G (*the kernel of* φ).

Let $H \subset G$ be a subgroup. For $x \in G$ the set $xH = \{xh : h \in H\}$ is called a *left coset of H*, and the set $Hx = \{hx : h \in H\}$ is called a *right coset of H*.

Lemma. *If $x, y \in G$, then the left cosets xH and yH are either identical or disjoint. In other words, the collection of distinct xH is a partition of G.*

PROOF: We check that the binary relation defined by "$x \in yH$" is an equivalence relation. The cosets xH are the elements of the corresponding partition.

a. Reflexive: $x \in xH$, since $x = xe$ and $e \in H$.

b. Symmetric: If $y \in xH$, then there is a $z \in H$ such that $y = xz$. This implies that $x = yz^{-1} \in yH$, since H is a subgroup.

c. Transitive: If $w \in yH$ and $y \in xH$, then for appropriate $h_1, h_2 \in H$, $y = xh_1$ and $w = yh_2 = xh_1 h_2$, and $w \in xH$ since $h_1 h_2 \in H$. ◀

The same proof shows that the set of *right* cosets of H in G forms a partition of G. The map $xH \mapsto Hx$ is clearly a (set-theoretic) bijection. The cardinality of the set of left (or right) cosets of H in G is called the *index* of H in G, and denoted by $[G : H]$.

A.3.5 Normal subgroups. Let G and G_1 be groups, and $\varphi \colon G \to G_1$ a homomorphism. Let $K \subset G$ be the kernel of φ, that is, the set of elements of G that are mapped by φ to e_1, the identity element of G_1. K is clearly a subgroup of G. We observe that if $\varphi(k) = e_1$ and $y \in G$ is arbitrary, then

$$\varphi(y^{-1}ky) = \varphi(y)^{-1}e_1\varphi(y) = e_1 \quad \text{and} \quad k = y^{-1}(yky^{-1})y,$$

so that $y^{-1}Ky = K$. In other words, K is mapped onto itself by every inner automorphism.

DEFINITION: A subgroup $H \subset G$ is a *normal subgroup* if it is mapped onto itself by every inner automorphism, that is, $H = y^{-1}Hy$ for every $y \in G$.

Let $H \subset G$ be a subgroup and let xH be a left coset of H; then $xH = xHx^{-1}x = H_1x$, which means that xH is a *right coset* of the group $H_1 = xHx^{-1}$. If H is a normal subgroup of G, then $H_1 = H$, and it follows that $xH = Hx$ for all $x \in G$.

We note this as a lemma:

Lemma. *If $H \subset G$ is normal, then every left coset of H is also a right coset.*

Proposition. *The quotient G/H of G by the equivalence relation defined by a normal subgroup H, has a natural group structure under which the map $x \mapsto xH$ is a homomorphism whose kernel is H.*

PROOF: Let xH and yH be cosets. Define $xH \cdot yH = xyH$. We need to show first that the product is well defined independently of the choice of the representatives x and y of the cosets. If we replace x by another representative xh_1, and y by yh_2, then, by the lemma, $h_1y = yh_3$ for some $h_3 \in H$, and we have $xh_1yh_2H = xyh_3h_2H = xyH$.

The facts that this multiplication is associative, that H is the unit element, and that the inverse of xH is $x^{-1}H$ are clear, and the fact that $x \mapsto xH$ is a homomorphism whose kernel is H is obvious. ◀

EXERCISES FOR SECTION A.3

exA.3.1 Check that, for any group G and every $y \in G$, the map $\varphi_y x = y^{-1}xy$ is an automorphism of G.

exA.3.2 Let G be a finite group of order m. Let $H \subset G$ be a subgroup. Prove that the order of H divides m.

exA.3.3 Let H be a normal subgroup of G and $U \subset G$ a subgroup.

a. Prove that $U \cap H$ is a normal subgroup of U.

b. Write $UH = \{uh : u \in U, h \in H\}$. Prove that UH is a subgroup of G.

c. Prove that UH/H is isomorphic to $U/(U \cap H)$.

$*$A.4   Group actions

A.4.1   Actions.

DEFINITION: An *action of G on a set X* is a homomorphism φ of G into $S(X)$, the group of invertible self-maps (permutations) of X.

The action defines a map $(g,x) \mapsto \varphi(g)x$. The notation $\varphi(g)x$ is often replaced by the simpler gx, when φ is implicitly understood. With the simpler notation, the assumption that φ is a homomorphism is equivalent to the conditions:

ga1. $ex = x$ for all $x \in X$ (e is the identity element of G).

ga2. $(g_1 g_2)x = g_1(g_2 x)$ for all $g_j \in G$, $x \in X$.

EXAMPLES:

a. G acts on itself ($X = G$) by left multiplication: $(x,y) \mapsto xy$.

b. G acts on itself ($X = G$) by right multiplication (by the inverse): $(x,y) \mapsto yx^{-1}$. (Remember that $(ab)^{-1} = b^{-1}a^{-1}$.)

c. G acts on itself by conjugation: $(x,y) \mapsto \varphi(x)y$ where $\varphi(x)y = xyx^{-1}$.

d. S_n acts as mappings on $[1,\ldots,n]$.

A.4.2 Orbits. The *orbit of an element $x \in X$ under the action of a group G* is the set $\mathrm{Orb}\,(x) = \{gx : g \in G\}$.

The orbits of a G-action form a partition of X. This means that any two orbits, $\mathrm{Orb}\,(x_1)$ and $\mathrm{Orb}\,(x_2)$, are either identical (as sets) or disjoint. In fact, if $x \in \mathrm{Orb}\,(y)$, then $x = g_0 y$ for some $g_0 \in G$, so that $y = g_0^{-1}x$, and $gy = gg_0^{-1}x$.

Since the set $\{gg_0^{-1} : g \in G\}$ is exactly G, we have $\mathrm{Orb}\,(y) = \mathrm{Orb}\,(x)$. If $\mathrm{Orb}\,(x_1) \cap \mathrm{Orb}\,(x_2)$ is not empty, take $\tilde{x} \in \mathrm{Orb}\,(x_1) \cap \mathrm{Orb}\,(x_2)$ and then $\mathrm{Orb}\,(\tilde{x}) = \mathrm{Orb}\,(x_1) = \mathrm{Orb}\,(x_2)$. It follows that the relation $x \equiv y$ defined by: $\mathrm{Orb}\,(x) = \mathrm{Orb}\,(y)$ is an equivalence relation on X.

EXAMPLES:

a. A subgroup $H \subset G$ acts on G by right multiplication: $(h,g) \mapsto gh$. The orbit of $g \in G$ under this action is the (left) coset gH.

b. S_n acts on $[1,\ldots,n]$, $(\sigma, j) \mapsto \sigma(j)$. Since the action is transitive, there is a unique orbit—$[1,\ldots,n]$.

c. If $\sigma \in S_n$, the group (σ) (generated by σ) is the subgroup $\{\sigma^k\}$ of all the powers of σ. Orbits of elements $a \in [1,\ldots,n]$ under the action of (σ), i.e., the sets $\{\sigma^k(a)\}$, are called *cycles* of σ and are written $(a_1,\ldots,a_l)$, where $a_{j+1} = \sigma(a_j)$, and l, the period of a_1 under σ, is the first positive integer such that $\sigma^l(a_1) = a_1$.

Notice that cycles are "enriched orbits", that is, orbits with some additional structure, here the cyclic order inherited from $\mathbb{Z}$. This cyclic order defines σ uniquely on the orbit, and is identified with the permutation that agrees with σ on the elements that appear in it, and leaves every other element in its place. For example, $(1,2,5)$ is the permutation that maps 1 to 2, maps 2 to 5, and 5 to 1, leaving every other element unchanged. Notice that n, the cardinality of the complete set on which S_n acts, does not enter the notation and is in fact irrelevant (provided that all the entries in the cycle are bounded by it; here $n \geq 5$). Thus, breaking $[1,\ldots,n]$ into σ-orbits amounts to writing σ as a product of disjoint cycles (see 4.1).

A.4.3   Conjugation. Two actions of a group G, $\varphi_1 : G \times X_1 \to X_1$, and $\varphi_2 : G \times X_2 \to X_2$ are *conjugate to each other* if there is an invertible map $\Psi : X_1 \to X_2$ such that for all $x \in G$ and $y \in X_1$,

(A.4.1) $\varphi_2(x)\Psi y = \Psi(\varphi_1(x)y)$ or, equivalently, $\varphi_2 = \Psi\varphi_1\Psi^{-1}$.

This is often stated as: *the following diagrams commute:*

$$
\begin{array}{ccc}
X_1 & \xrightarrow{\varphi_1} & X_1 \\
\downarrow{\scriptstyle\Psi} & & \downarrow{\scriptstyle\Psi} \\
X_2 & \xrightarrow{\varphi_2} & X_2
\end{array}
\quad\text{or, equivalently,}\quad
\begin{array}{ccc}
X_1 & \xrightarrow{\varphi_1} & X_1 \\
\uparrow{\scriptstyle\Psi^{-1}} & & \downarrow{\scriptstyle\Psi} \\
X_2 & \xrightarrow{\varphi_2} & X_2
\end{array}
$$

meaning that the composition of maps associated with arrows along a path depends only on the starting and the end point, and not on the path chosen.

A.5   Rings and algebras

A.5.1   Rings.

DEFINITION: A *ring* is a triplet $(\mathscr{R}, +, \cdot)$, where $\mathscr{R}$ is a set, and $+$ and $\cdot$ are binary operations on $\mathscr{R}$ called addition and multiplication respectively, such that $(\mathscr{R}, +)$ is a commutative group, the multiplication is associative (but not necessarily commutative), and the addition and multiplication are related by the *distributive laws*:

$$a(b+c) = ab + ac \quad\text{and}\quad (b+c)a = ba + ca.$$

A *subring* $\mathscr{R}_1$ of a ring $\mathscr{R}$ is a subset of $\mathscr{R}$ that is a ring under the operations induced by the ring operations, i.e., addition and multiplication, in $\mathscr{R}$.

A commutative ring with multiplicative identity is often called a *domain*. A domain R in which $ab = 0$ implies $a = 0$ or $b = 0$ is called an *integral domain*.

EXAMPLES:

a.   Any field, $\mathbb{F}$, is also a ring.

b. $\mathbb{Z}$ is an integral domain.

c. For $q \in \mathbb{Z}$, $q > 1$, the subring $q\mathbb{Z} = \{qn : n \in \mathbb{Z}\}$ is a commutative ring *without* multiplicative identity.

d. The (finite) ring $\mathbb{Z}_q$, defined in **ex1.1.4**, is a field if q is prime. If $q = nm$, with $n, m > 1$, then $\mathbb{Z}_q$ is an example of a domain that is *not* an integral domain, since n and m are nonzero and $nm = 0$ in $\mathbb{Z}_q$.

e. $\mathscr{M}(n)$ with multiplication defined by matrix multiplication, is a ring with identity that is not commutative.

A.5.2 Ideals.

DEFINITION: A *left (resp. right) ideal* in a ring $\mathscr{R}$ is a subring I that is closed under multiplication on the left (resp. right) by elements of $\mathscr{R}$: for $a \in \mathscr{R}$ and $h \in I$ we have $ah \in I$ (resp. $ha \in I$). A *two-sided ideal* is a subring that is both a left ideal and a right ideal.

If the ring is commutative, the adjectives "left", "right" are irrelevant.

Assume that $\mathscr{R}$ has an identity element. For $g \in \mathscr{R}$, the set $I_g = \{ag : a \in \mathscr{R}\}$ is a left ideal in $\mathscr{R}$, and is clearly the smallest (left) ideal that contains g.

Ideals of the form I_g are called *principal left ideals*, and g is called a *generator* of I_g. One defines *principal right ideals* similarly.

A.5.3 Principal ideal domains.

DEFINITION: An integral domain in which every ideal is principal is called a *principal ideal domain*, (P.I.D.).

Principal ideal domains are important in number theory in the study of divisibility and factorization. The canonical example of a P.I.D. is $\mathbb{Z}$. The proof that $\mathbb{Z}$ is a P.I.D. follows from "division with remainder".

Theorem (Division with remainder in $\mathbb{Z}$). *For every pair $m, n \in \mathbb{Z}$, with $m, n \neq 0$, there exists a unique pair $q, r \in \mathbb{Z}$ with $0 \leq r < |n|$, such that*

(A.5.1) $$m = qn + r.$$

PROOF: We prove the existence, and leave uniqueness as an exercise.

If $m = qn$, with $q \in \mathbb{Z}$, then the result is obvious, so we assume that n does not divide m.

Now, assume that $m, n > 0$, and let r_0 be the smallest *positive* integer in the set $S = \{m - qn : q \in \mathbb{Z}\}$. If $r \geq n$, then $r_1 = r_0 - n$ is a smaller nonnegative member of S, contradicting the minimality of r_0, so $0 \leq r_0 < n$. If $q_0 = (m - r_0)/n$, then the pair q_0, r_0 is as required.

If $m > 0$ and $n < 0$, and q_0, r_0 is the pair that works for m and $|n|$, then $-q_0$ and r_0 work for m and n.

If $m < 0$ and $n > 0$, and q_0, r_0 is the pair that works for $|m|$ and n, then $-(q_0 + 1)$ and $n - r_0$ work for m and n.

Finally, if $m, n < 0$, and q_0, r_0 is the pair that works for $|m|$ and $|n|$, then $q_0 + 1$ and $n - r_0$ work for m and n. ◀

Corollary. $\mathbb{Z}$ *is a principal ideal domain.*

PROOF: Let m be the smallest positive element of I and $n \in I$, $n > 0$. By the theorem above, we can divide with remainder: $n = qm + r$ with q, r integers, and $0 \leq r < m$. Since both n and qm are in I, so is r. Since m is the smallest positive element in I, $r = 0$ and $n = qm$. Thus, all the positive elements of I are divisible by m (and so are their negatives). ◀

If $m_j \in \mathbb{Z}$, $j = 1, 2$, the set $I_{m_1, m_2} = \{n_1 m_1 + n_2 m_2 : n_1, n_2 \in \mathbb{Z}\}$ is an ideal in $\mathbb{Z}$, and hence has the form $g\mathbb{Z}$. As g divides every element in I_{m_1, m_2}, it divides both m_1 and m_2; as $g = n_1 m_1 + n_2 m_2$ for appropriate n_j, every common divisor of m_1 and m_2 divides g. It follows that g is their *greatest common divisor*, $g = \gcd(m_1, m_2)$. We summarize:

Proposition. *If m_1 and m_2 are integers, then for appropriate integers n_1, n_2,*

$$\gcd(m_1, m_2) = n_1 m_1 + n_2 m_2.$$

A.5.4 Euclidean domains. The proof that $\mathbb{Z}$ is a P.I.D. can be repeated almost verbatim for any ring that has an appropriate notion of division with remainder.

DEFINITION: An integral domain R is called a *Euclidean domain* if there exists a function v from the set of nonzero elements in R to the nonnegative integers that satisfies the *division with remainder property* in R: if $a, b \in R$ and $b \neq 0$, then there exist $q, r \in R$, with $v(r) < v(b)$ or $r = 0$, such that

(A.5.2) $$a = qb + r.$$

Such a function is called a *valuation* on R. We note that in general, (i.e., in Euclidean domains other than $\mathbb{Z}$), the pair q, r is not necessarily unique.

Theorem. *If R is a Euclidean domain, then R is a principal ideal domain.*

PROOF: Let I be a nontrivial ideal in R, and let $a \in I$ be an element of minimal value in I, i.e., $v(a) \leq v(b)$ for all $b \neq 0$ in I. Then the proof that a is a generator of I is identical to the proof of Corollary A.5.3. ◀

The most important example of a Euclidean domain (besides $\mathbb{Z}$) is the ring of polynomials over a field (see A.6).

A.5.5 Fields. Fields were defined in Chapter 1. We observe that a field may be defined equivalently as a domain in which every nonzero element is invertible. The most basic examples are the fields of rational numbers, $\mathbb{Q}$; real numbers, $\mathbb{R}$; and complex numbers, $\mathbb{C}$, which are all *infinite* fields.

As mentioned in Chapter 1, there are also finite fields. In exercise **ex1.1.4**, we show how to construct a field with exactly p elements, where p is a prime number. More generally, it can be shown that for every prime p and every integer $k \geq 1$, there is exactly one field (up to *isomorphism*) with p^k elements. See example *d* below.

Two fields, $\mathbb{F}_1$ and $\mathbb{F}_2$, are *isomorphic* if there is a bijective map $\varphi : \mathbb{F}_1 \to \mathbb{F}_2$ satisfying $\varphi(a+b) = \varphi(a) + \varphi(b)$ and $\varphi(ab) = \varphi(a)\varphi(b)$, for all $a, b \in \mathbb{F}_1$.

DEFINITION: If $\mathbb{F} \subset \mathbb{K}$ are fields, then we say that $\mathbb{F}$ is a *subfield* of $\mathbb{K}$, and equivalently, that $\mathbb{K}$ is an *extension* of $\mathbb{F}$. More generally, $\mathbb{K}$ is an extension of $\mathbb{F}$ if $\mathbb{K}$ contains a subfield that is isomorphic to $\mathbb{F}$.

An extension $\mathbb{K}$ of $\mathbb{F}$ is also a *vector space* over $\mathbb{F}$. The degree of the field extension $\mathbb{K}$ of $\mathbb{F}$ is, by definition, the dimension of $\mathbb{K}$ as a vector space over $\mathbb{F}$. The extension is *finite* if its degree is finite.

EXAMPLES:

a. $\mathbb{C}$ is an extension of $\mathbb{R}$ of degree 2.

b. $\mathbb{R}$ is an extension of $\mathbb{Q}$ of infinite degree.

c. $\mathbb{Q}(\sqrt[3]{2}) = \{a + b\sqrt[3]{2} + c\sqrt[3]{4} : a,b,c \in \mathbb{Q}\}$ is an extension of $\mathbb{Q}$ of degree 3.

d. $\mathbb{F}_4 = \{0,1,a,b\}$, with addition and multiplication given in the tables below, is a degree 2 extension of $\mathbb{Z}_2$.

+	0	1	a	b
0	0	1	a	b
1	1	0	b	a
a	a	b	0	1
b	b	a	1	0

×	0	1	a	b
0	0	0	0	0
1	0	1	a	b
a	0	a	b	1
b	0	b	1	a

A.5.6 Algebras.

DEFINITION: An *algebra* over a field $\mathbb{F}$ is a ring $\mathscr{A}$ and a multiplication of elements of $\mathscr{A}$ by scalars (elements of $\mathbb{F}$), that is, a map $\mathbb{F} \times \mathscr{A} \to \mathscr{A}$ such that if we denote the image of (a,u) by au, we have, for $a,b \in \mathbb{F}$ and $u,v \in \mathscr{A}$,

$$\text{identity:} \quad 1u = u;$$
$$\text{associativity:} \quad a(bu) = (ab)u, \quad a(uv) = (au)v;$$
$$\text{distributivity:} \quad (a+b)u = au + bu \quad \text{and} \quad a(u+v) = au + av.$$

A *subalgebra* $\mathscr{A}_1 \subset \mathscr{A}$ is a subring of $\mathscr{A}$ that is also closed under multiplication by scalars.

A left (resp. right, resp. two-sided) ideal in an algebra $\mathscr{A}$ is a subalgebra of $\mathscr{A}$ that is closed under left (resp. right, resp, either left or right) multiplication by elements of $\mathscr{A}$.

Examples:

a. $\mathbb{F}[x]$, the algebra of polynomials in one variable x with coefficients from $\mathbb{F}$, and the standard addition, multiplication, and multiplication by scalars. It is an algebra over $\mathbb{F}$.

b. $\mathbb{C}[x,y]$, the (algebra of) polynomials in two variables x, y with complex coefficients, and the standard operations. $\mathbb{C}[x,y]$ is a "complex algebra", that is, an algebra over $\mathbb{C}$.

Notice that by restricting the scalar field to, say, $\mathbb{R}$, a complex algebra can be viewed as a "real algebra", i.e., an algebra over $\mathbb{R}$. The underlying field is part of the definition of an algebra. The "complex" and the "real" $\mathbb{C}[x,y]$ are *different algebras*.

c. $\mathscr{M}(n)$, the $n \times n$ matrices with matrix multiplication as product.

EXERCISES FOR SECTION A.5

exA.5.1 Let $\mathscr{R}$ be a ring with identity, and $B \subset \mathscr{R}$ a set. Prove that *the ideal generated by B*, that is, the smallest ideal I in $\mathscr{R}$ that contains B, is the set $I = \{\sum a_j b_j : a_j \in \mathscr{R}, b_j \in B\}$.

exA.5.2 Verify that $\mathbb{Z}_p$ is a field.
Hint: If p is a prime and $0 < m < p$ then $\gcd(m, p) = 1$.

exA.5.3 Prove that the set of invertible elements in a ring with an identity is a multiplicative group.

exA.5.4 Show that the set of polynomials $\{P : P = \sum_{j \geq 2} a_j x^j\}$ is an ideal in $\mathbb{F}[x]$, and that $\{P : P = \sum_{j \leq 7} a_j x^j\}$ is an additive subgroup but not an ideal.

A.6   Polynomials

Let $\mathbb{F}$ be a field and $\mathbb{F}[x]$ the algebra of polynomials $P = \sum_0^n a_j x^j$ in the variable x with coefficients from $\mathbb{F}$. The *degree* of P, $\deg(P)$, is the highest power of x appearing in P with non-zero coefficient. If $\deg(P) = n$, then $a_n x^n$ is called the *leading term* of P, and a_n the *leading coefficient*. A polynomial is called *monic* if its leading coefficient is 1.

A.6.1   Division with remainder. By definition, an ideal in a ring is *principal* if it consists of all the multiples of one of its elements, called a *generator* of the ideal. The ring $\mathbb{F}[x]$ shares with $\mathbb{Z}$ the property of being a *principal ideal domain*—every ideal is principal. The proof for $\mathbb{F}[x]$ is virtually the same as the one we had for $\mathbb{Z}$, and is again based on *division with remainder*.

Theorem. *Let $P, F \in \mathbb{F}[x]$. There exist polynomials $Q, R \in \mathbb{F}[x]$ such that $\deg(R) < \deg(F)$ and*

$$(A.6.1) \qquad\qquad P = QF + R.$$

PROOF: Write $P = \sum_{j=0}^{n} a_j x^j$ and $F = \sum_{j=0}^{m} b_j x^j$ with both a_n and b_m nonzero, so that $\deg(P) = n$ and $\deg(F) = m$.

If $n < m$, there is nothing to prove: $P = 0 \cdot F + P$.

If $n \geq m$, we write $q_{n-m} = a_n/b_m$, and $P_1 = P - q_{n-m}x^{n-m}F$, so that $P = q_{n-m}x^{n-m}F + P_1$ with $n_1 = \deg(P_1) < n$.

If $n_1 < m$, we are done. If $n_1 \geq m$, write the leading term of P_1 as $a_{1,n_1}x^{n_1}$, and set $q_{n_1-m} = a_{1,n_1}/b_m$, and $P_2 = P_1 - q_{n_1-m}x^{n_1-m}F$. Now $\deg(P_2) < \deg(P_1)$ and $P = (q_{n-m}x^{n-m} + q_{n_1-m}x^{n_1-m})F + P_2$.

Repeating the procedure a total of k times, $k \leq n - m + 1$, we obtain $P = QF + P_k$ with $\deg(P_k) < m$, and the statement follows with $R = P_k$. ◄

A useful application of the theorem is the observation that a polynomial $P \in \mathbb{F}[x]$ vanishes at a point $\lambda \in \mathbb{F}$ if and only if P is divisible by $x - \lambda$. Division with remainder gives $P = (x - \lambda)Q + r$, with $r \in \mathbb{F}$. Evaluating at $x = \lambda$ shows that the remainder r is equal to $P(\lambda)$.

The same idea in a more abstract context gives the following:

Corollary. *Let $I \subset \mathbb{F}[x]$ be an ideal, and let P_0 be an element of minimal degree in I. Then P_0 is a generator for I.*

PROOF: If $P \in I$, write $P = QP_0 + R$, with $\deg(R) < \deg(P_0)$. Since $R = P - QP_0 \in I$, and 0 is the only element of I whose degree is smaller than $\deg(P_0)$, we have $P = QP_0$. ◄

Since a generator of I divides every P in I, it is an element of minimal degree in I, so that a polynomial in I is a generator *if and only if* it is an element of minimal degree in I.

The generator P_0 is unique up to multiplication by a scalar. If P_1 is another generator, each of the two divides the other, and since they have the same degree, the quotients are scalars. It follows that if we normalize P_0 by requiring that it be *monic*, that is, with leading coefficient 1, it is unique and we refer to it as *the* generator.

A.6.2 Given polynomials P_j, $j = 1, \ldots, l$, any ideal that contains them all must contain all the polynomials $P = \sum q_j P_j$ with arbitrary polynomial coefficients q_j. On the other hand, the set of all theses sums is clearly an ideal in $\mathbb{F}[x]$. It follows that the ideal generated by $\{P_j\}$ is equal to the set of polynomials of the form $P = \sum q_j P_j$ with polynomial coefficients q_j.

The generator G of this ideal divides every one of the P_j's, and, since G can be expressed as $\sum q_j P_j$, every common factor of all the P_j's divides G. In other words, $G = \gcd\{P_1, \ldots, P_l\}$, the greatest common divisor of $\{P_j\}$. This implies

Theorem. *Given polynomials P_j, $j = 1, \ldots, l$, there exist polynomials q_j such that $\gcd\{P_1, \ldots, P_l\} = \sum q_j P_j$.*

In particular:

Corollary. *If P_1 and P_2 are relatively prime, there exist polynomials q_1, q_2 such that $P_1 q_1 + P_2 q_2 = 1$.*

A.6.3 Factorization. A polynomial P in $\mathbb{F}[x]$ is *irreducible* or *prime* if it has no proper factors, that is, if every factor of P is either scalar multiple of P or a scalar.

Lemma. *If $\gcd(P, P_1) = 1$ and $P \mid P_1 P_2$, then $P \mid P_2$.*

PROOF: There exist q, q_1 such that $qP + q_1 P_1 = 1$. Then

$$qPP_2 + q_1 P_1 P_2 = P_2,$$

the left-hand side is divisible by P and hence so is P_2. ◄

An immediate extension of the lemma gives that if an irreducible polynomial P divides a product of polynomials, then it divides (at least) one of the factors.

Corollary. *Assume*

(A.6.2)
$$\prod_{j=1}^{n} P_j = \prod_{k=1}^{m} \Phi_k$$

with all the factors irreducible and, for simplicity, monic. Then the sets of factors $\{\Phi_k\}$ and $\{P_j\}$, including repetitions, are identical.

In fact, every P_j divides (at least) one of the Φ_k's and is therefore equal to one of them. Eliminating equal factors on both sides exhausts both sets of factors.

Theorem (Prime power factorization). *Every $P \in \mathbb{F}[x]$ admits a factorization $P = \prod \Phi_j^{m_j}$, where each factor Φ_j is irreducible in $\mathbb{F}[x]$, and they are all distinct.*

The factorization is unique up to the order in which the factors are enumerated, and up to multiplication by non-zero scalars.

PROOF: The uniqueness was established in the corollary above. We prove the existence of the factorization by induction on $\deg(P)$. A linear polynomial P is irreducible and has the trivial factorization: $P = P$. Assume that the statement is valid for all $k < n$ and let $P \in \mathbb{F}[x]$ of degree n. If P is irreducible, the factorization is again trivial, $P = P$. If P is reducible, $P = P_1 P_2$ where both factors have positive degrees k_1 and k_2 respectively, and $k_1 + k_2 = n$. Since $k_j < n$, we can apply the induction hypothesis, write $P_1 = \prod_j \Phi_j^{l_j}$ and $P_2 = \prod_j \Psi_j^{m_j}$, and combine the two factorizations to a factorization for P.   ◄

A.6.4   A polynomial $P \in \mathbb{F}[x]$ *splits* over $\mathbb{F}$ if P is a product of linear factors in $\mathbb{F}[x]$. As we observed in A.6.1, $(x - \lambda)$ divides P if and only if λ is a root of P, i.e., if $P(\lambda) = 0$. Thus, finding the linear factors of P is the same as finding the roots of P.

Lemma. *If $\Psi \in \mathbb{F}[x]$ is irreducible, then there is a finite extension of $\mathbb{F}$ in which Ψ has a root.*

PROOF: Let A be the companion matrix of Ψ. From the discussion in 5.3.3, it follows that $\Psi = \min P_A$. Since Ψ is irreducible, the algebra $\mathscr{P}(A) = \{P(A) : P \in \mathbb{F}[x]\}$ is a field (Proposition 5.3.6). The subfield $\{aI : a \in \mathbb{F}\}$ of $\mathscr{P}(A)$ is isomorphic to $\mathbb{F}$, so $\mathscr{P}(A)$ is a finite extension of $\mathbb{F}$ (of degree $= \deg\Psi$). Since $\Psi(A) = \min P_A(A) = 0$, Ψ has a root in $\mathscr{P}(A)$. ◀

If $\Phi \in \mathbb{F}[x]$ does not split over $\mathbb{F}$, then Φ has an irreducible, non-linear factor $\Psi_1 \in \mathbb{F}[x]$. By the lemma, there is a finite extension $\mathbb{K}_1$ of $\mathbb{F}$ in which Ψ_1 has a root. If Φ does not split over $\mathbb{K}_1$, then Φ has an irreducible, nonlinear factor $\Psi_2 \in \mathbb{K}_1[x]$. Applying the lemma again, we find a finite extension $\mathbb{K}_2$ of $\mathbb{K}_1$ in which Ψ_2 has a root. Repeating this process no more than $\deg\Phi$ times proves the following basic result from field theory.

Theorem. *If $\Phi \in \mathbb{F}[x]$, then $\mathbb{F}$ has a finite extension $\mathbb{K}$ such that Φ splits in $\mathbb{K}[x]$.*

EXAMPLES:

a. The polynomial $x^2 + 1$ is irreducible in $\mathbb{R}[x]$ and splits in $\mathbb{C}[x]$.

b. The polynomial $x^2 + x + 1$ is irreducible in $\mathbb{Z}_2[x]$ and splits in $\mathbb{F}_4[x]$.

A.6.5 The fundamental theorem of algebra.

DEFINITION: A field $\mathbb{F}$ is *algebraically closed* if every $P \in \mathbb{F}[x]$ has roots in $\mathbb{F}$, that is, elements $\lambda \in \mathbb{F}$ such that $P(\lambda) = 0$.

The *fundamental theorem of algebra* states that $\mathbb{C}$ is algebraically closed.

Theorem. *Given a non-constant polynomial P with complex coefficients, there exists a complex number λ such that $P(\lambda) = 0$.*

A.6.6 It is an immediate consequence of the definitions that $\mathbb{F}$ is algebraically closed if and only if every polynomial in $\mathbb{F}[x]$ splits in $\mathbb{F}[x]$, or equivalently, if and only if the only irreducible (prime) polynomials in $\mathbb{F}[x]$ are linear.

Thus, over an algebraically closed field, the prime-power factorization theorem takes the form:

Theorem. *Let $\mathbb{F}$ be algebraically closed and let $P \in \mathbb{F}[x]$ be a polynomial of degree n. There exist $\lambda_1, \ldots, \lambda_n \in \mathbb{F}$ (not necessarily distinct) and a $\neq 0$ (the leading coefficient of P) such that*

(A.6.3) $$P(x) = a \prod_{1}^{n} (x - \lambda_j).$$

A.6.7 Factorization in $\mathbb{R}[x]$. The factorization (A.6.3) applies, of course, to polynomials with real coefficients, but the roots need not be real. The basic example is $P(x) = x^2 + 1$ with the roots $\pm i$.

We observe that for polynomials P whose coefficients are all real, we have $P(\bar{\lambda}) = \overline{P(\lambda)}$, which means in particular that if λ is a root of P, then so is $\bar{\lambda}$.

A second observation is that

(A.6.4) $$(x - \lambda)(x - \bar{\lambda}) = x^2 - 2x\Re\lambda + |\lambda|^2$$

has real coefficients.

Combining these observations with (A.6.3) we obtain that the prime factors in $\mathbb{R}[x]$ are linear polynomials and quadratic polynomials of the form (A.6.4) where $\lambda \notin \mathbb{R}$.

Theorem. *Let $P \in \mathbb{R}[x]$ be a polynomial of degree n. P admits a factorization*

(A.6.5) $$P(z) = a \prod (x - \lambda_j) \prod Q_j(x),$$

where a is the leading coefficient, $\{\lambda_j\}$ is the set of real zeros of P and Q_j are irreducible quadratic polynomials of the form (A.6.4) corresponding to (pairs of conjugate) nonreal roots of P.

Either product may be empty, in which case it is interpreted as 1.

As mentioned above, the factors appearing in (A.6.5) need not be distinct—the same factor may be repeated several times. We can rewrite the product as

$$(A.6.6) \qquad P(z) = a \prod_j (x - \lambda_j)^{l_j} \prod_j Q_j^{k_j}(x),$$

with λ_j and Q_j now distinct, and the exponents l_j resp. k_j their multiplicities. The factors $(x - \lambda_j)^{l_j}$ and $Q_j^{k_j}(x)$ appearing in (A.6.6) are pairwise relatively prime.

A.6.8 The symmetric functions theorem.

DEFINITION: A polynomial in m variables $P(x_1, \dots, x_m)$ is *symmetric* if, for any permutation $\tau \in S_m$,

$$(A.6.7) \qquad P(x_{\tau(1)}, \dots, x_{\tau(m)}) = P(x_1, \dots, x_m).$$

EXAMPLES:

a. $s_k = s_k(x_1, \dots, x_m) = \sum_j^m x_j^k.$

b. $\sigma_k = \sigma_k(x_1, \dots, x_m) = \sum_{i_1 < \dots < i_k} x_{i_1} \cdots x_{i_k}.$ We assume that $k \leq m$ in this case.

The *Symmetric Functions Theorem* states that every symmetric polynomial in m variables of degree $\leq k$ is a polynomial, with rational coefficients, in $\{s_1, \dots, s_k\}$. For our purposes, (see 5.1.3), the following special case is sufficient.

Proposition. $\sigma_k(x_1, \dots, x_m)$ *is a polynomial with rational coefficients in the functions* $s_1(x_1, \dots, x_m), \dots, s_k(x_1, \dots, x_m)$.

PROOF: Observe that $\sigma_1 = s_1$, and $s_1^2 = s_2 + 2\sigma_2$, so $\sigma_2 = \frac{1}{2}(s_1^2 - s_2)$. In general, we observe that $s_1^k - s_k - k! \sigma_k$ is a polynomial (with integer coefficients) in $\{s_1, \dots, s_{k-1}\}$ and $\{\sigma_1, \dots, \sigma_{k-1}\}$, and the statement follows by induction. ◀

Corollary. *Let P be a monic polynomial of degree n in $\mathbb{F}[x]$, and let $\{\lambda_1, \ldots, \lambda_n\}$ be the roots of P, repeated according to their multiplicity, and lying (perhaps) in some finite extension of $\mathbb{F}$. Then the coefficients of P are polynomials with rational coefficients in*

$$s_1(\lambda_1, \ldots, \lambda_n), \ldots, s_{n-1}(\lambda_1, \ldots, \lambda_n).$$

PROOF: We have

$$P(x) = (x - \lambda_1) \cdots (x - \lambda_n) = \sum_{j=0}^{n} (-1)^{n-j} \sigma_{n-j}(\lambda_1, \ldots, \lambda_n) x^j,$$

and the statement follows from the proposition above.   ◀

*A.6.9   **Continuous dependence of the zeros of a polynomial on its coefficients.** Let $P(z) = z^n + \sum_0^{n-1} a_j z^j$ be a monic polynomial and let $r > \sum_0^n |a_j| = 1 + \sum_0^{n-1} |a_j|$. If $|z| \geq r$, then $|z|^n > |\sum_0^{n-1} a_j z^j|$, so that $P(z) \neq 0$. All the zeros of P are located in the disc $\{z : |z| \leq r\}$.

Denote $E = \{\lambda_k\} = \{z : P(z) = 0\}$, the set of zeros of P, and, for $\eta > 0$, denote by E_η the "η-neighborhood" of E, that is, the set of points whose distance from E is less than η.

The key remark is that $|P(z)|$ is bounded away from zero in the complement of E_η for every $\eta > 0$. In other words, given $\eta > 0$, there is a positive ε such that the set $\{z : |P(z)| \leq \varepsilon\}$ is contained in E_η.

Proposition. *With the preceding notation, given $\eta > 0$, there exists $\delta > 0$ such that if $P_1(z) = \sum_0^n b_j z^j$ and $|a_j - b_j| < \delta$, then all the zeros of P_1 are contained in E_δ.*

PROOF: If $|P(z)| > \varepsilon$ on the complement of E_η, take δ small enough to guarantee that $|P_1(z) - P(z)| < \frac{\varepsilon}{2}$ in $|z| \leq 2r$.   ◀

Corollary. *Let $A \in \mathcal{M}(n, \mathbb{C})$ and $\eta > 0$ be given. There exists $\delta > 0$ such that if $A_1 \in \mathcal{M}(n, \mathbb{C})$ has all its entries within δ from the corresponding entries of A, then the spectrum of A_1 is contained in the η-neighborhood of the spectrum of A.*

PROOF: The closeness condition on the entries of A_1 to those of A implies that the coefficients of χ_{A_1} are within δ' from those of χ_A, and δ' can be guaranteed to be arbitrarily small by taking δ small enough.

◀

Index

Symbols

Titles in This Series

TITLES IN THIS SERIES